STUDENT UNIT

AS Geography
UNIT 1

Specification A

Unit 1: Physical Environments

Andy Palmer

AS Geography

Philip Allan Updates
Market Place
Deddington
Oxfordshire
OX15 0SE

tel: 01869 338652
fax: 01869 337590
e-mail: sales@philipallan.co.uk
www.philipallan.co.uk

© Philip Allan Updates 2002

ISBN 0 86003 691 X

All rights reserved; no part of this publication may be reproduced, stored in a retrieval system, or transmitted, in any form or by any means, electronic, mechanical, photocopying, recording or otherwise without either the prior written permission of Philip Allan Updates or a licence permitting restricted copying in the United Kingdom issued by the Copyright Licensing Agency Ltd, 90 Tottenham Court Road, London W1P 9HE.

This Guide has been written specifically to support students preparing for the Edexcel Specification A AS Geography Unit 1 examination. The content has been neither approved nor endorsed by Edexcel and remains the sole responsibility of the author.

Printed by Raithby, Lawrence & Co. Ltd, Leicester

Edexcel (A) Unit 1

Contents

Introduction
About this guide 4
AS Geography 4
Revision advice 6
Examination skills 6

Content Guidance
About this section 12
Earth systems
The earth's crust 13
Igneous activity 17
Weathering 20
Fluvial environments
The hydrological cycle 24
Rivers 26
River processes 29
Coastal environments
Coastal processes 32
Coastal landforms 34
Coastal ecosystems 36

Questions and Answers
About this section 40
Q1 Earth systems (I) 41
Q2 Earth systems (II) 45
Q3 Fluvial environments (I) 49
Q4 Fluvial environments (II) 53
Q5 Coastal environments (I) 57
Q6 Coastal environments (II) 61

AS Geography

Introduction

About this guide

This guide is for students following the Edexcel Specification A AS Geography course. It aims to guide you through Unit Test 1, which examines the content of **Unit 1: Physical Environments**.

This guide will clarify:
- the content of the unit so that you know and understand what you have to learn
- the nature of the unit test
- the geographical skills and techniques that you will need to know for the assessment
- the standards you will need to reach to achieve a particular grade
- the examination techniques you will require to improve your performance and maximise your achievement

This **Introduction** describes the structure of AS Geography and outlines the aims and method of examining Unit 1. It then provides advice on learning and revision techniques before explaining some of the key command words used in examination papers. There is also advice concerning general and specific geographical skills.

The **Content Guidance** section summarises the essential information of Unit 1. It is designed to make you aware of the material that has to be covered and learnt. In particular, the meaning of key terms is made clear.

The **Question and Answer** section provides sample questions and candidate responses at C-grade level and A-grade level. Each answer is followed by a detailed examiner's response. It is suggested that you read through the relevant topic area in the Content Guidance section before attempting a question from the Question and Answer section, and only read the specimen answers and examiner's comments after you have tackled the question yourself.

AS Geography

AS is designed to be an intermediate standard between GCSE and A-level. While it is difficult to specify exactly what that standard is, it recognises that students will only have followed the course for one year.

After you have completed the AS course, you may decide to stop studying the subject. Alternatively, you can continue with the A2 course, which has different units and is designed to be more demanding. This will ultimately enable you to combine your AS results with your A2 results and to receive an A-level grade.

Scheme of assessment

Unit 1: Physical Environments is one of three units that make up the AS specification. It is assessed by a written paper which is marked out of 60 marks and worth 90 uniform marks (i.e. 30%) of the 300 marks that make up the whole assessment. The AS marks make up 50% of the total A-level assessment. The three A2 units make up the other 50%.

Unit	Unit exam length	Max mark	Max uniform mark	AS weighting
1: Physical Environments	1 hour 15 minutes	60	90	30%
2: Human Environments	1 hour 15 minutes	60	90	30%
3: Fieldwork Investigation	1 hour 30 minutes or Personal Enquiry	60	120	40%

Unit 1

The specification content of Unit 1 comprises three sections:
- Earth systems
- Fluvial environments
- Coastal environments

These are studied at a range of scales from local (such as individual landforms) to global (such as the pattern of tectonic plates). There is a need to appreciate the interaction of human activity with the physical environment and there is a requirement to study specific places and examples.

During the course you should aim to:
- develop a knowledge of geographical terminology, concepts, principles and theories
- acquire and apply knowledge and understanding of physical processes, their interactions and outcomes over space and time, through the study of places and environments
- acquire and apply a range of geographical and transferable skills necessary for the study of geography, particularly to describe, analyse and interpret data and resources
- develop an understanding of the relationships between physical environments and people
- appreciate the dynamic nature of geography — how places and environments change

Systems

One of the most useful concepts that underlies much of the specification content is the systems approach. Its key feature is that it recognises the existence of three elements which are linked:

Inputs ⟶ Throughputs ⟶ Outputs

There are two commonly used forms of the approach:
- It can be used to look at the transfer of a substance, for example, water through the drainage basin system.

Inputs	→	Throughputs	→	Outputs
e.g. precipitation		e.g. infiltration		e.g. discharge

- It can be used to look conceptually at the shaping of the landscape.

Inputs	→	Throughputs	→	Outputs
Controlling factors	→	Processes	→	Geographical features
e.g. wave energy		e.g. hydraulic action		e.g. arch

If you are able to explain how wave energy is linked to hydraulic action and how hydraulic action contributes to the formation of an arch, then you have a good grasp of the topic!

Revision advice

You will probably find that you have a busy year if you are studying four or even five AS subjects! It is, therefore, really important that you keep up to date with your studying and complete all work set on time. If you are absent for any reason, you need to find out what you missed and catch up as quickly as possible. Hopefully you will have some time for revision at the end of the course and you should try to make full use of any opportunities that you are given. When you are revising on your own, it is generally found to be helpful if you do the following:

- Plan a revision schedule that gives fairly even time allocation to each of your subjects and to each unit within the subject.
- Stick to the schedule!
- Read through all material relating to a topic — notes, handouts, questions, worksheets etc.
- Make revision notes that summarise the key elements of the topic (the Content Guidance section of this book will help).
- Make a list of the key facts/figures for each of the main case studies/examples that you need.
- Practise sample/past examination questions (such as the ones in the Question and Answer section).
- Ask your teacher/lecturer to clarify any topics/issues about which you are uncertain.
- Remember, you need to revise *all* of the specification content in this unit.

Examination skills

For each topic, you will be assessed on your knowledge, understanding and skills. The structure of the examination paper is such that you have to show that you have studied *all* of the topics in the specification content — you cannot pick and choose!

Edexcel (A) Unit 1

In the examination paper, there will be *six* structured data response-type questions. Two questions will be set on each of the three sub-sections. You must answer *three* questions in 1 hour 15 minutes, choosing *one* from each section. Each answer will be marked out of 20, giving a total of 60 marks.

Choice of questions

The questions in the paper will be in the same order as in the specification: questions 1 and 2 will be on earth systems; questions 3 and 4 on fluvial environments; and questions 5 and 6 on coastal environments.

Always look at the whole of each question in a section before you make your choice. The resource at the start may be familiar, indeed it may be from the textbook that you use, but can you answer all parts of the question? In particular, it is worth looking at the final part as this will often be worth 6 marks and may require knowledge of a located example. Do you have one that is appropriate? Do you have some locational detail, i.e. names of places, facts and figures?

Answering the question

- *Study the stimulus material carefully and be prepared to use it in the first couple of parts.*
- *Look through the whole question as this can help you avoid repeating yourself.*
- *Plan your answer before you start writing.*
- *Keep your answers concise and to the point.* You need to answer the questions in the spaces provided on the question paper. The amount of space should indicate how much you need to write. Don't worry if you run out of lines! If there is some blank space underneath, you can finish your answer there, or ask the invigilator for a supplementary answer sheet if you need more space. Don't forget to tie this to your paper before you hand it in and make sure you indicate which question you are answering. However, if you keep writing more than the lines suggest is necessary, you may be drifting off the point of the question and you may run out of time!
- *Make sure you understand what the question is asking you to do.* The command word is particularly important. The list below gives the meanings of command words often used in this paper.
  - **Describe**: give details of the appearance and characteristics
  - **Explain**: give reasons for
  - **Outline**: provide general principles or main features without great detail
  - **Suggest**: offer possible reasons that logically might be appropriate
  - **How**: identify the process or mechanism
  - **Why**: explain
  - **Define**: give the meaning of
  - **State**: give a very brief, possibly one-word, answer
- *When you are asked to refer to a located example, you should not only state the name/location at the start, but also ensure that you write about the example in the answer.* A good way of doing this is to provide a couple of facts and figures that relate specifically to the location.

- *If you are drawing a diagram or map, don't worry about trying to make it a work of art!* It should be neat, clear and well-labelled or annotated, as required. A label identifies something while an annotation can provide explanation or comment.

Skills

In each question, there are a few marks available for the demonstration of competence in skills. At AS, most of the skill marks are available in Unit 3, so you should not expect to have to undertake lengthy statistical calculations in this unit. However, you may have to work out the mean or range of a set of data, for example. Skills also include description and this will often be required. For instance, you may be asked to describe the relationship between two sets of data on a graph or the distribution of a variable on a map.

Three steps can usually be followed in order to describe successfully:
- recognise the general trend or pattern
- use evidence from the resource
- spot any anomalies or exceptions

You should also be familiar with the use of a wide range of **stimulus resources**:
- sections of text and extracts from articles
- OS maps, especially at 1:50 000 and 1:25 000
- choropleth and isopleth maps
- aerial and satellite images
- annotated sketches and photographs
- annotated sketch maps
- line graphs
- bar graphs
- pie and divided line graphs
- scattergraphs and best-fit lines
- triangular graphs
- flow lines
- sketch sections
- proportional symbols

You will encounter lots of these during your course, in books, tests and in practical questions. Make sure you know what they are used for, how to extract information from them and how to add further information to them.

You may also be asked to complete a simple **statistical calculation**, including:
- mean, mode, median
- quartiles and inter-quartile ranges
- Spearman's rank-order coefficient
- chi-squared test

In each case, you should be able not only to undertake the calculation, but also to interpret the result.

Timing

You have roughly 25 minutes to answer each 20-mark question. You should find that this is plenty of time and that you will be able to:
- spend a few minutes choosing which question you are going to answer in each section
- stop and think when a question is proving a challenge
- plan your answer to the final part of the question, which is worth the most marks
- check through your answers at the end

If you are really stuck on part of a question, leave it and carry on; you can always come back to it later with a fresh view.

Quality of written communication

There are no separate marks added on at the end for this, but it is assessed by the mark scheme that the examiners use when marking your paper. It is particularly important in the last part of each question where you are writing at greater length. You should aim to:
- spell, punctuate and apply rules of grammar
- structure and order your answers in a logical manner
- use appropriate geographical terminology

It can be useful to compile a list of key terms and their meanings for each topic. Many such terms appear in bold in the Content Guidance section of this book.

Common examination mistakes

Examiners are often saddened and frustrated by the apparent inability of candidates to perform to their potential in written examinations. There are some very common mistakes that candidates make which lead to the following comments from the examiners:

'Lack of focus'
- Not responding correctly to the command word.
- Not recognising the specific demands of the question.
- Including irrelevant material just because it has been learned.

'Out of time'
Candidates often spend too long on one question because:
- they are struggling with one of the sub-sections and are stopping to think for too long, instead of moving on and coming back later
- they know a lot about the topic and are writing at great length, even though much of what they are writing is irrelevant

'Poorly located'
The final part of each question will often require the use of a located example. Candidates must have an appropriate example to use and show their located

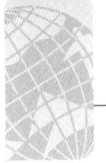

knowledge by referring to names of specific places and giving facts and figures. This might include rainfall data, names of rock types, names of tributaries of the main river, etc.

'Gaps in knowledge'

It is a requirement that at AS, students need to cover *all* of the specification content. You cannot concentrate on selected elements of a topic and leave other elements out. For example, you may be strong on plate tectonics but weak on weathering in the earth systems section. You cannot neglect weathering, because both question 1 and question 2 might have a sub-section on this topic.

There are three sections in the specification content for Unit 1:

(1) Earth systems

(2) Fluvial environments

(3) Coastal environments

In this Content Guidance, each of the three elements will be considered in terms of:
- the key concepts involved
- the content required
- the use of examples and case studies

This should make it clear to you what you need to know and understand.

It should be noted that the synoptic links in each element are not dealt with in this guide. Although these may be taught as part of the AS course, they are not assessed until A2.

Earth systems

The earth's crust

Evidence for continental drift and plate tectonics
There is a range of evidence that can be used in support of the theories of continental drift and plate tectonics:

- the jigsaw fit of continents, particularly South America and Africa, especially when the edge of the continental shelf is used rather than the present coastline
- fossils, such as mesosaurus, a small, freshwater reptile found in both Africa and South America
- the age of basaltic rocks on the sea floor. Near the mid-Atlantic ridge the basalts are much younger than those found towards the continental shelf
- palaeomagnetism of the basalts, which suggests that rocks were formed at the ridge and then moved away from the ridge in both directions. There is a parallel pattern of magnetism in the rocks
- similar rock types and geological structures on either side of the Atlantic, for example the Appalachian mountains of North America and the Caledonian mountains of Scotland
- coal and oil reserves in Antarctica and glacial striations in India, which suggest that these areas were once in different climatic zones

Do remember that **continental drift** refers to the original idea that the continental land masses have changed position over time, without the mechanism being understood, whilst **plate tectonic theory** recognises that the plates are the units of movement and that a valid mechanism for this movement has been suggested.

> **Exam tip** Make sure that you are able to relate each piece of evidence to the theory. For example, the oldest basalts on the sea floor are found furthest away from the mid-oceanic ridges. This suggests that they were formed at the ridge and have subsequently moved away to be replaced by younger rocks.

The global pattern of plates, their direction and rate of movement
Figure 1 on page 14 shows all the information that you need. You are not expected to know the rates of movement for each plate, but you should have some appreciation of the scale of these movements and be aware that the rates vary significantly from plate to plate.

Possible mechanisms of movement, including convection currents
You should be aware that the most likely cause of plate movement is the existence of convection cell currents in the **asthenosphere**, caused by radioactive decay in the **core**. This causes the **lithosphere** to be moved by the circulating **magma**.

> **Exam tip** The convection mechanism is not fully understood; you should phrase any answers tentatively, for example 'it is thought that...' or 'it is very likely that...'

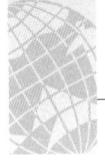

AS Geography

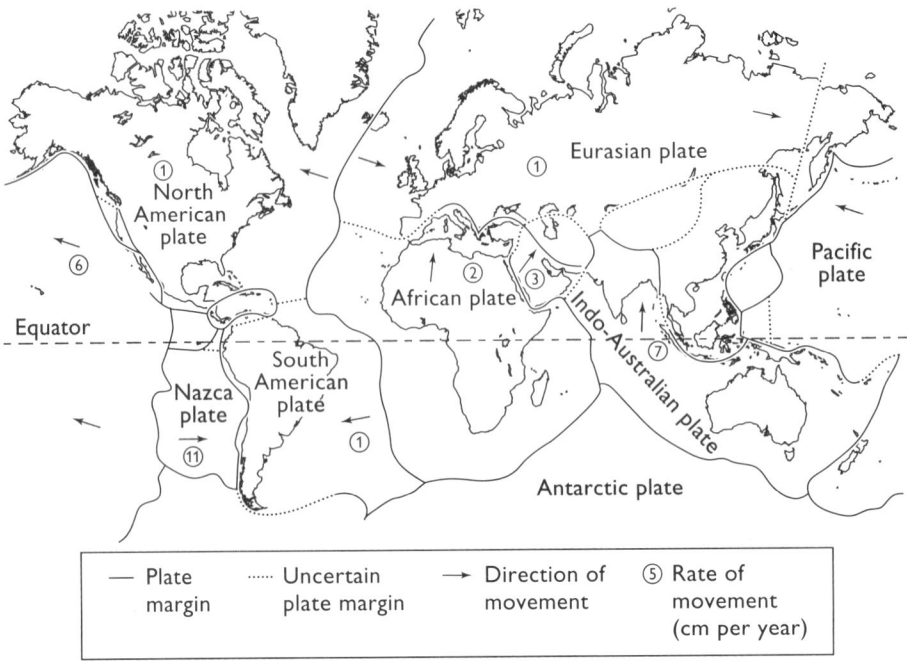

Figure 1 The global pattern of plates

Processes at constructive, destructive and conservative plate margins

Processes at plate margins include **subduction, convergence, divergence** and **earthquake activity**.

Do ensure that you are familiar with this terminology as different books (and specifications!) tend to use different terms.

> **Case study tip** You *must* be able to refer to a specific example of each type of margin. Some examples are listed below.
> - Constructive: mid-Atlantic — American and Eurasian/African plates diverging.
> - Destructive: South American and Nazca plates converging.
> - Conservative: Pacific and North American plates moving laterally in the same direction but at different rates.
>
> It is particularly helpful to be able to draw a cross-sectional diagram through the margin example to show the *processes* that are taking place and the *features* that result. Examples are shown in Figure 2.

> **Exam tip** If you are drawing a diagram, *customise* it to fit the demands of the question. For example, if the question asks for landforms, concentrate on including a good range of landforms on the diagram. If the question asks for more than one thing, perhaps processes and landforms, consider showing each in a different colour on your diagram. Not only does it make it clearer, it also provides a good focus on the question.

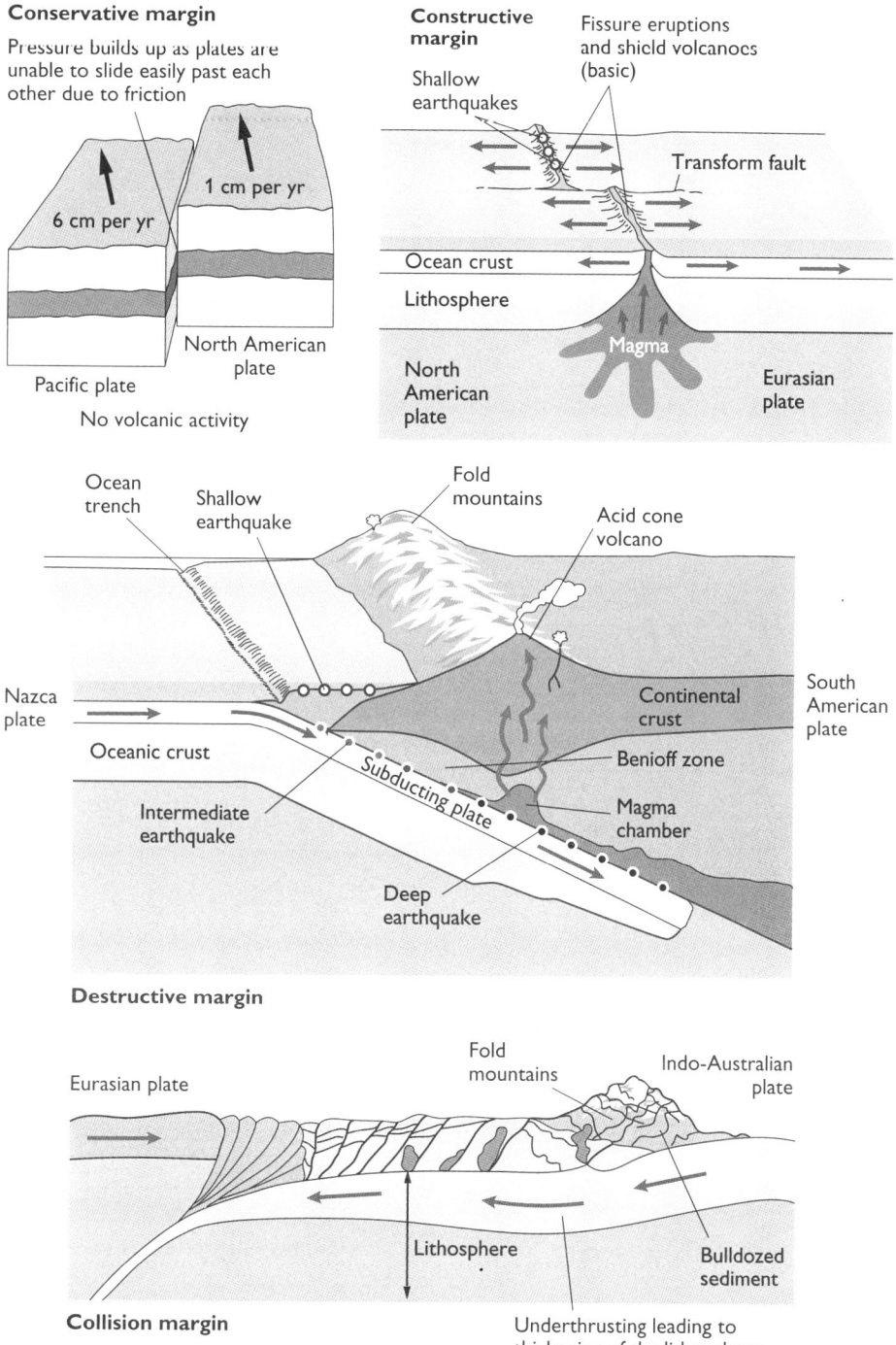

Figure 2 Processes at plate margins

You should also be aware of the results of convergence between two continental plates. For example, convergence between the Indo-Australian and Eurasian plates leads to the formation of fold mountains (Himalayas). Convergence between two oceanic plates, such as the Pacific and the Philippines plates, can lead to an island arc — in this case, the Mariana Islands.

Earthquake activity

Earthquake activity occurs at all plate boundaries. An earthquake is a movement or shaking of the ground caused by the release of pressure which has built up as plates move in relation to each other, hindered by friction between them. The nature of earthquake activity varies with the type of plate margin.

Margin type	Earthquake activity
Constructive	• Mainly shallow focus, with high frequency and low magnitude • Plates diverge, causing pressure to be readily released near the surface
Destructive	• Variable depth of focus, from shallow at the ocean trench to deep (600 km) in the subduction zone • Generally high magnitude, but low frequency, due to the huge build-up of pressure as the plates converge
Conservative	• Shallow focus as plates 'slide' past each other • Some high-magnitude, low-frequency activity when pressure does build up (e.g. San Francisco 1906/1989), but also much that is low magnitude, high frequency (e.g. four minor earthquakes on the San Andreas fault on 26 January 2000)

Hot spots

Hot spots are places where plumes of magma are rising from the asthenosphere, even though they are not necessarily near a plate margin. If the crust is particularly thin or weak, the magma may escape onto the surface as a volcanic eruption. Lava can build up over time until it is above present-day sea level, giving rise to a volcanic island, for example Mauna Loa, Hawaii.

Resulting global pattern of landforms

Feature	Examples	Type of plate margin
Fold mountains	Himalayas, Rockies, Andes, Southern Alps	Destructive/collision
Ocean trench	Peru–Chile, Marianas, Aleutian	Destructive
Island arc	Philippine Islands, Japan, Mariana Islands	Destructive
Ocean ridge	Mid-Atlantic, Carlsberg, East-Pacific	Constructive

You need to be able to *describe* the general pattern of these features as well as know the names of major examples. You should also be able to link the feature to the appropriate type of plate margin as a way of being able to explain their distribution. Figure 3 shows the global distribution of tectonic landforms.

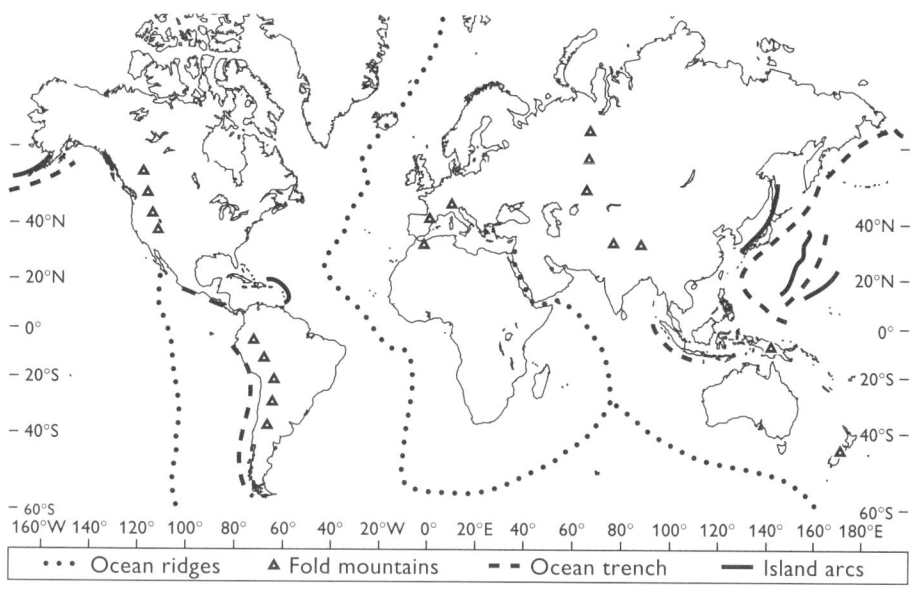

Figure 3 Global distribution of tectonic landforms

Igneous activity

Distribution of volcanic activity

You should be able to *describe* the global distribution as shown in Figure 4.

Figure 4 Global distribution of volcanoes

Volcanoes occur in a series of broad bands. These tend to be either along the edge of continental land masses, for example the west coast of South America (e.g. Nevada del Ruiz), or through the middle of oceans, for example mid-Atlantic (e.g. Heimaey). There are exceptions to this general pattern. For instance, the Hawaiian Islands (e.g. Mauna Loa) are more isolated and not in a broad band.

You should also be able to *explain* the distribution by linking the areas of volcanic activity to constructive and destructive plate margins and hot spots. There are some significant differences between the volcanoes at constructive and destructive margins.

Margin type	Characteristics
Constructive	• Fissure eruptions or shield volcanoes with long, gently-sloping sides • Lava tends to be basic, fluid and hot (1200°C), with a low silica content • Eruptions are generally low magnitude but high frequency
Destructive	• Acid cone or composite volcanoes with short, steep sides • Lava is acid, viscous and cooler (800°C), with a higher silica content • Eruptions tend to be high magnitude but low frequency

Exam tip If you are explaining the distribution, remember to mention that there is no volcanic activity at conservative margins as no new crust is being created or existing crust being destroyed.

Characteristics and formation of extrusive landforms

Extrusive landform	Characteristics	Formation	Impact on the landscape
Lava plateau	Extensive layers of basaltic lava	Basaltic lava, usually from a fissure, spreading over a wide area; a series of lava and flows builds the plateau in layers	Deccan Plateau, India is 700 000 km² and made from 29 flows; it is flat, featureless and extensive
Acid cone	Steep-sided; acid lava — high silica content	Destructive margin eruption; viscous lava cools quickly, without flowing far before solidifying	Mount Pelée is 1400 m high, rising steeply from the surrounding surface
Basic cone	Gently sloping; basic lava — low silica content	Hot spot eruption; low viscosity lava which flows a long way before cooling and solidifying	Mauna Loa, Hawaii has slopes of 6° and a diameter of 500 km at the sea floor
Composite cone	Alternating layers of ash and lava	After a lava eruption, the lava solidifies in the vent, causing a build-up of pressure; the next eruption is violent and fragments the cone into ash	Mount Etna is 5° near the base and 3° near the summit; secondary (parasitic) cones develop on the side when the main vent is blocked
Fissure	Lava erupted along a line of weakness	At constructive margins, basic lava escapes along fault lines/fractures; it spreads long distances	Mid-Atlantic has persistent outpourings which produce the gently sloping sides of the ridge

> **Case study tip** You should be able to write about *at least one* landscape that has been produced by extrusive volcanic activity. The Hawaiian Islands and the mid-Atlantic ridge are good examples. It may be helpful to have studied a range of landscapes, not all of which need to be in great detail, in order to illustrate differences.

Characteristics and formation of intrusive landforms

All three features listed in the table below are formed by magma rising towards the surface, but cooling and solidifying before being extruded onto the surface. The magma cools slowly as it is not exposed to the air, and so mineral crystals (e.g. quartz in granite) grow to a large size.

Batholiths can cause a general doming up of the surface, for example the Isle of Arran, but they are only exposed after the gradual weathering and erosion of the less resistant, overlying 'country rock'.

The heat and/or pressure exerted on the country rock causes **metamorphic** rock to be produced, for example:

sandstone ⟶ gneiss, limestone ⟶ marble

Intrusive landform	Characteristics	Formation	Impact on the landscape
Dyke	A vertical intrusion	Magma cuts across bedding planes in the country rock, following lines of weakness (e.g. joints)	At Kildonan, Arran, a series of dykes 2–3 m wide is exposed across the beach
Sill	A horizontal intrusion	Magma flows along the bedding planes which provide a line of weakness	At Drumadoon, Arran, a sill has been exposed, forming a 50 m high cliff
Batholith	A dome-shaped mass of igneous rock	Rising magma pushes up in the country rock	The whole of the Isle of Arran was domed up by the intrusion of a batholith; the overlying sandstone has now been removed

> **Case study tip** You need to be able to write about *at least one* landscape that has been affected by intrusive activity. The Isle of Arran is a very good example; Dartmoor could also be used.

Economic benefits of igneous activity

You need to be aware that there is a range of benefits that igneous activity may bring. These are outlined in the table overleaf.

> **Exam tip** Although specific examples are not required in the assessment, they can be a useful way of illustrating and supporting an idea.

Economic benefit	Illustration
Building materials	Granite — Aberdeen is known as 'Granite City' as so many buildings are made from it
Minerals	Tin — used to be extracted from mines in Devon and Cornwall, from the batholiths of Dartmoor and Bodmin Moor
Geothermal energy	Iceland — most of Reykjavik's domestic heating is supplied by hot water from a geothermal plant at Nesjavellir
Tourism	Isle of Arran — the dykes and sills attract visitors, who create a demand for tertiary employment and increased spending in the local economy

Weathering

Chemical, physical and biological weathering

Weathering is defined as *the breakdown and decay of rocks by the elements of the weather (except wind) acting in situ*. It is frequently confused with **erosion** by AS students. Erosion involves wearing away by moving forces, such as running water, breaking waves, advancing glaciers and the wind.

Breakdown is largely achieved by **physical weathering** processes which produce smaller fragments of the same rock.

Decay is the result of **chemical weathering**, which involves chemical reactions between the elements of the weather and some minerals within the rock. It produces a weak residue of different material.

Biological weathering may consist of physical actions (such as the growth of plant roots) or chemical processes (such as chelation by organic acids).

You *must* know how each of the processes listed in the table below operates. You might also know about others, such as hydrolysis, particularly as you need to have studied a granite landscape in another section of this unit.

Type of weathering	Processes involved	What happens?
Chemical	Oxidation	Iron in rocks reacts with oxygen in the air or in water; it becomes soluble under extremely acidic conditions
	Carbonation	Rain water combines with dissolved carbon dioxide from the atmosphere, to produce weak carbonic acid; this reacts with calcium carbonate in rocks such as limestone to produce calcium bicarbonate, which is soluble
	Solution	Some salts are soluble in water; other minerals, such as iron, are only soluble in very acidic water
	Hydrolysis	Feldspar in granite reacts with hydrogen in water to produce kaolin (china clay)

Type of weathering	Processes involved	What happens?
Physical	Freeze–thaw	Water enters cracks/joints and expands by nearly 10% when it freezes; this exerts pressure on the rock, causing pieces to break off
	Pressure release	When overlying rocks are removed by weathering and erosion, the underlying rock expands and fractures parallel to the surface; this is significant in the exposure of granite batholiths
	Thermal expansion	Rocks expand when heated and contract when cooled; if they are subjected to frequent cycles of temperature change, the outer layers may crack and flake off
Biological	Tree roots	Tree roots grow into cracks/joints and exert outward pressure, having a similar effect to freeze–thaw
	Organic acids	Organic acids, produced during decomposition of litter, cause soil water to become more acidic and react with some minerals, in a process called chelation

Case study tip There is no requirement to provide any case study detail of these processes. However, you need to link the processes to granite and limestone landscapes.

Physical, human and temporal factors influencing the rate of weathering
Climate

Types and rates of weathering vary with climate, in particular moisture availability and temperature. This is illustrated in Peltier's diagram (Figure 5).

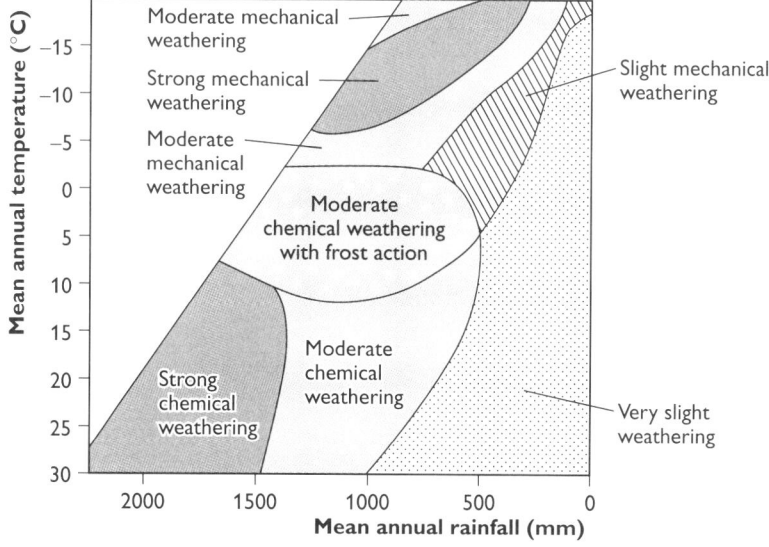

Figure 5 Peltier's diagram of weathering

Moisture is important as it is *the medium for most chemical reactions* and it is present in most physical processes. Temperature is important, as most chemical reactions take place faster at higher temperatures. Temperature fluctuations are also significant; the number of freeze–thaw cycles influences freeze–thaw weathering, and the diurnal (daily) range affects thermal expansion.

Geology
Some rock types are susceptible to specific processes because of their mineral composition. For example, limestone contains calcium carbonate and is prone to carbonation; granite suffers from hydrolysis due to the presence of feldspar. The existence of joints, cracks, bedding planes and pores permits water to enter the rock and increases the surface area being weathered by water-based processes (both physical and chemical).

Relief
This can affect microclimates, cause water to collect at the base of slopes or encourage downslope movement of weathered debris, exposing fresh rock beneath to the elements.

Soil/vegetation cover
This tends to increase rates of chemical weathering by increasing the acidity of rain water draining through. However, it decreases rates of physical weathering by protecting the rock from extremes of temperature.

Human activity
Atmospheric pollution can increase the acidity of rain water. Vegetation cover may be added/removed (see above). Global warming may increase rates of chemical weathering.

Time
Rates may vary over time as the influence of other factors changes. Rocks might be gradually weakened by weathering over time and eventually weather more rapidly.

You should realise that in any location, a range of different weathering processes might be acting and that a combination of different factors will control their rates.

The impact of weathering on the landscape

Regolith is the collective term for all weathered material. It may be fragments of rock from physical processes or chemical residues from chemical processes. **Scree** is angular fragments of weathered rock, often accumulating at the base of a slope. **Soil** is partly formed from the input of minerals from weathering.

Granite landscapes are usually upland areas, partly due to the general resistance of granite to weathering and erosion, and partly as it causes the surface to become domed upwards when it is intruded as a batholith.

One of the most distinctive features is the **tor**. There are differing theories about tor formation, but a possible sequence of events is:
- intrusion of a granite batholith beneath the surface

- deep chemical weathering by hydrolysis, most rapid where a closely spaced pattern of cooling joints exists
- removal of overlying rocks by surface weathering, erosion and possibly solifluction (periglacial mass movement)
- present-day weathering (e.g. freeze–thaw) further enlarges pressure release joints
- chemically weathered debris (e.g. kaolin) is removed to the surrounding valley bottoms
- physically weathered debris (e.g. blockfields and clitter) accumulates around the tor
- the tor remains exposed — an isolated mass of bare rock, up to 20 m high, consisting of a number of individual boulders resting on top of the core stones beneath

Dartmoor has a number of good examples, such as Hay Tor and Yes Tor.

Case study tip You are required to study a particular granite landscape and you should be able to name and locate specific features within the area. The Isle of Arran and Dartmoor are good examples.

Limestone landscapes are distinctive on account of their permeability and susceptibility to carbonation weathering. Their most striking feature is the **limestone pavement**, a large area of bare, exposed limestone. The following processes lead to its formation:

- glacial action removes soil and vegetation cover, exposing the highly jointed limestone
- water enters the joints and enlarges them by carbonation to form **grykes**
- the blocks between the joints remain upstanding as **clints**
- water may sit on the surface of the clints and form small depressions, called **solution hollows**
- as this water runs off into the grykes, it forms small grooves called **karren**
- vegetation may colonise the grykes, increasing weathering rates by producing organic acids
- freeze–thaw also exploits the joints, leading to the formation of **scree**
- **dolines** (or shake holes) are surface depressions, caused by the collapse of underlying limestone which has been severely affected by carbonation, perhaps due to close joint spacing

Case study tip You are required to study a particular limestone landscape and you should be able to name and locate specific features within the area. The area above Malham in the Yorkshire Dales is a good example, with scree slopes at Goredale Scar nearby. The Burren in Western Ireland is a huge limestone pavement.

The impact of weathering on human activity

You need to be aware that there is a range of both positive and negative impacts of weathering.

Impact	Example
Building damage	St Paul's Cathedral has suffered rapid weathering, particularly on the south side facing Bankside Power Station (now closed); expensive repairs were needed
Scenic value	Malham attracts visitors who create a demand for tertiary employment and increase spending in the local economy
China clay extraction	Jobs and trade have been created at Lee Moor in Cornwall
Transport	Netting has been applied to bare limestone cuttings alongside the M5 in the Gordano Valley near Bristol, to stop scree falling on the motorway

Exam tip Although specific examples will not be required in the assessment, they can be a useful way of illustrating and supporting an idea.

Fluvial environments

The hydrological cycle

The global hydrological cycle

The **global hydrological cycle** is a *closed* system in that it does not have external inputs and outputs (see Figure 6). There is a fixed amount of water in the earth–atmosphere system, which can exist in different states (liquid, gas and solid); the *proportion* of water that exists in each state can change over time.

Figure 6 The global hydrological cycle

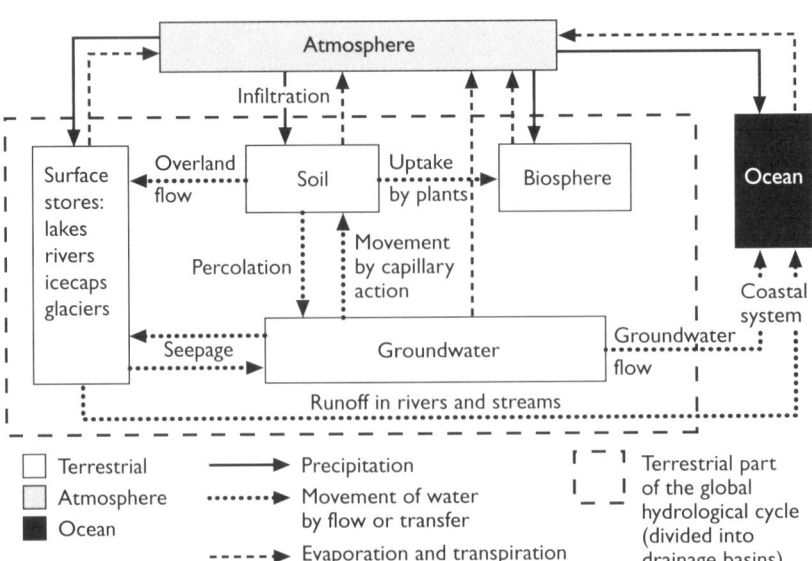

The drainage basin cycle

The **drainage basin cycle** is an *open* system in that it has external inputs and outputs and the amount of water in the basin varies over time.

- **Evapotranspiration** is the diffusion of water from vegetation and water surfaces into the atmosphere.
- **Condensation** is the process by which water vapour changes into water droplets when air becomes saturated at its dew point temperature.
- **Precipitation** includes all deposits of moisture on the earth's surface from the atmosphere (dew, hail, rain, snow, etc.).
- **Surface runoff** (or overland flow) occurs either when the rate of rainfall exceeds the rate of infiltration or when the soil is saturated.
- **Groundwater flow** is water moving underground through pores, joints, etc.
- **Evaporation** is the process by which water droplets change into water vapour.
- **Transpiration** involves the evaporation of moisture through the pores on leaf surfaces.
- **Infiltration** is the movement of water from the surface into the soil.
- **Percolation** is the movement of water from the soil into the underlying geology.

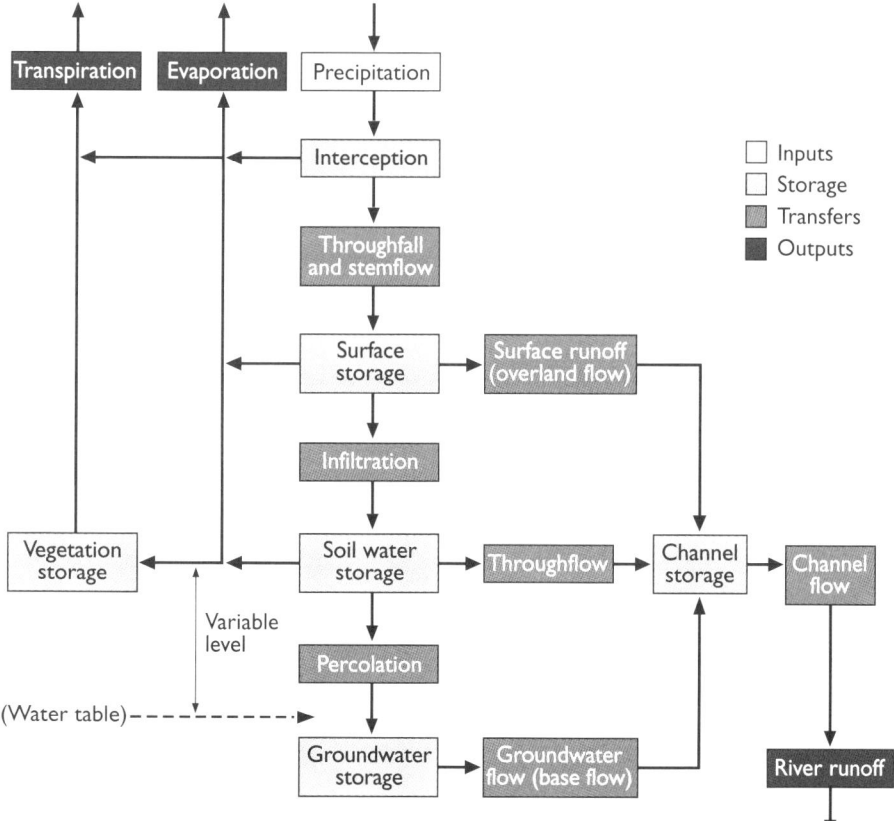

Figure 7 The drainage basin cycle

The causes of rainfall

The key to cloud formation is that air rises, cools and becomes saturated, leading to condensation. This can be caused by:

- air being forced to rise over a relief barrier, such as a hill (**orographic**)
- warm air rising over cold air, as the warm air is less dense than the cold air (**frontal**)
- air rising because it has been heated by the ground, which has absorbed solar radiation, and become less dense (**convectional**)

Tiny particles in the air act as condensation nuclei, around which condensation takes place to produce water droplets. These then increase in size, possibly by colliding with other droplets, until they are heavy enough to fall through rising air as raindrops.

> **Exam tip** You do not need to know about the other forms of precipitation, such as snow, hail, dew, etc.

Rivers

Characteristics of river regimes and the physical and human factors influencing them

The **regime** of a river is its pattern of discharge, usually over the period of a year. The **discharge** is the volume of water passing a point on a river over a period of time. It is calculated by:

cross-sectional area × velocity

and is usually expressed in cumecs or cubic metres per second.

> **Case study tip** You need to have studied two rivers with contrasting regimes. For each, you should be able to *describe* the regime (peaks, troughs, seasonal variations, etc.) and *explain* the physical and human factors that influence its shape. These are likely to include:
> - physical factors—rainfall pattern, snow melt, temperature (evaporation), relief, geology
> - human factors — dams, water supply, irrigation

> **Exam tip** Be careful not to confuse river regimes with storm hydrographs, even though the same factors can be applied.

Characteristics of hydrographs and the physical and human factors influencing them

A **storm hydrograph** is the changing discharge of a river over time in response to a specific input of precipitation. The main features of a typical hydrograph are labelled in Figure 8.

The shape of a hydrograph may vary over time on the same river (temporal variation) and also from one river to another (spatial variation). The factors that influence the shape of a hydrograph are shown in the table below. It should be appreciated that these vary in importance and that they also interact with each other.

Factor	Short lag time, high peak, steep rising limb ('flashy')	Long lag time, low peak, gently-sloping rising limb
Weather/climate	Intense storm Rapid snow melt Low evaporation rates	Steady rainfall Slow snow melt High evaporation rates
Rock type	Impermeable	Permeable
Soil	Clays	Sands
Relief	Steep	Gentle/flat
Vegetation	Bare/low-density Deciduous in winter	Dense Deciduous in summer
Antecedent conditions	Basin already wet from previous rain, water table high, soils saturated	Basin dry, water table low, soils unsaturated
Human activity	Urbanisation Deforestation	Low population density Afforestation

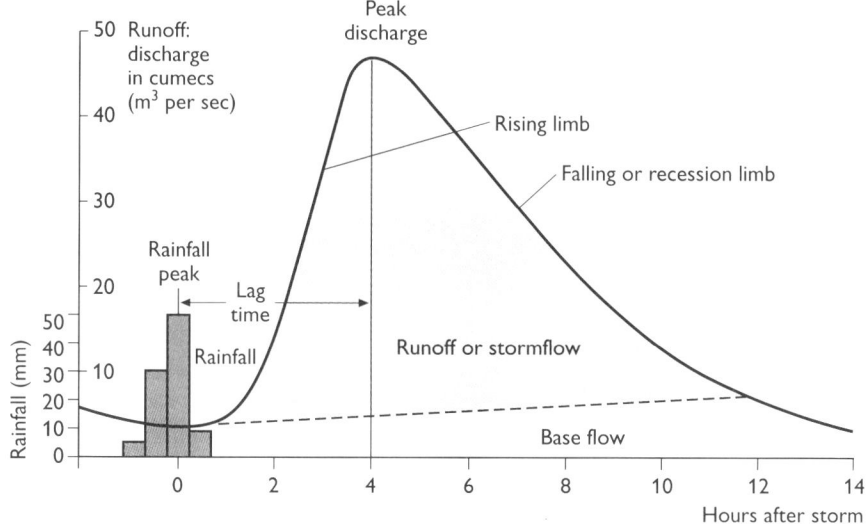

Figure 8 Features of a typical hydrograph

Case study tip You do not need to have studied the hydrograph of a particular river/storm, but you need to be able to apply your knowledge of the factors to an unfamiliar example.

Physical and human causes of flooding

Rivers flood when they exceed their bankfull capacity and water leaves the channel to flow over the surrounding area. The causes of flooding are varied and generally the result of the combined effects of several factors. These factors are very likely to include those shown above as leading to a 'flashy' hydrograph (which often precedes flooding).

Case study tip You need to have a good knowledge of the causes of a particular flood event, such as Mississippi, 1993 or West Midlands, 1998. You should be able to provide facts, figures and located detail.

Downstream changes in velocity, discharge, efficiency (hydraulic radius) and channel shape

Theoretically, the variables listed above should change with distance downstream, as shown in Bradshaw's model (see Figure 9).

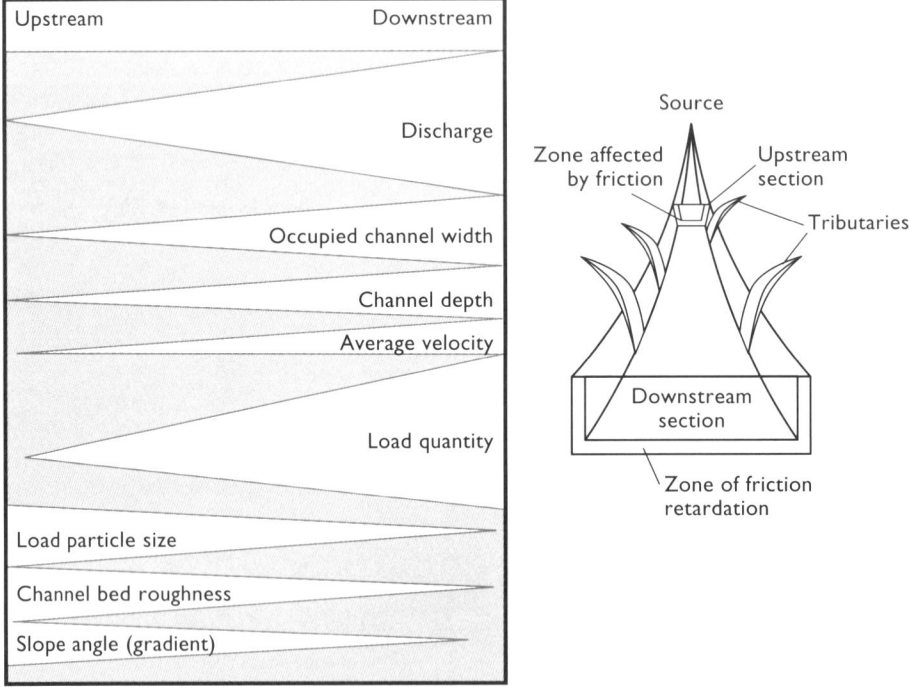

Figure 9 Bradshaw's model of how channel variables change downstream

Velocity is influenced by gradient, channel roughness and efficiency. It increases downstream, despite the fact that gradient normally decreases. This suggests that the other factors, which are both related to friction, are more important.

Discharge increases downstream as more water is added to the channel by tributaries and groundwater flows. As velocity is also a component of discharge, it is affected by the higher velocities downstream.

Hydraulic radius is calculated by:

$$\frac{\text{cross-sectional area}}{\text{wetted perimeter}}$$

Channel shape tends to become more efficient (indicated by a higher hydraulic radius value) with distance downstream. This means that, *proportionally*, there is less water in contact with the frictional effects of the bed and banks (see Figure 10).

Channel A

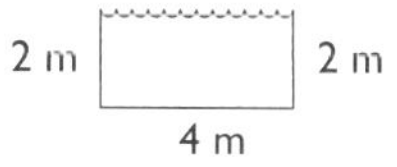

2 m │ 2 m
    4 m

Cross-sectional area = 2 × 4 = 8 m²
Wetted perimeter = 2 + 4 + 2 = 8 m
Hydraulic radius = $\frac{CSA}{WP} = \frac{8}{8} = 1\,m^2/m$

Channel B

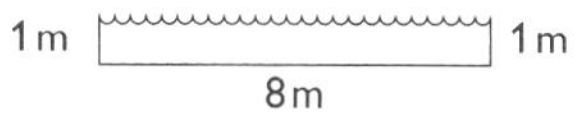

1 m │ 1 m
    8 m

Cross-sectional area = 1 × 8 = 8 m²
Wetted perimeter = 1 + 8 + 1 = 10 m
Hydraulic radius = $\frac{CSA}{WP} = \frac{8}{10} = 0.8\,m^2/m$

Channel A has a higher hydraulic radius and is therefore more efficient

Figure 10 How the efficiency of channel shape increases with distance downstream

Case study tip You need to have investigated the downstream changes in these variables on a particular river. This allows you to compare the actual changes with those expected and to explain why they differ. This may well be done by undertaking fieldwork on a local river.

River processes

The channel processes of erosion, transportation and deposition

Some specific mechanisms of the basic river processes are given in the table below.

Process	Mechanism	What happens
Erosion	Abrasion	The load of the river rubs against the bed/banks and wears them away
	Attrition	The particles of load become worn away by colliding with each other and as they rub against the bed/banks
	Corrosion	Weakly acidic river water reacts with some minerals in the rock of the channel bed/banks (e.g. calcium carbonate in limestone)
	Cavitation	The cumulative effect of pressure imparted by the collapse of tiny bubbles in the water
Transportation	Suspension	Small particles, such as clay, are held in the body of moving water
	Solution	Dissolved material, such as calcium, is carried in the water
	Saltation	Larger particles, such as small stones, are repeatedly lifted and then dropped, as they are too heavy to be carried continually in suspension
	Traction	Large rocks and boulders are pushed along the bed, particularly in times of flood
Deposition	Loss of energy due to a decrease in velocity	Velocity decreases on the inside bend of a meander when the river enters the sea or a lake after a flood
	Loss of energy due to a decrease in volume	Volume decreases during a drought, when evaporation rates are high, or when flowing over permeable bed material

Note that material from river deposition is characterised by being smooth/rounded and sorted sequentially by size (the largest material being deposited first).

This sub-topic includes the resulting landforms such as valleys, waterfalls, rapids, meanders, braids, levees, oxbow lakes, deltas and flood plains.

> **Exam tip** For each of the landforms listed above, you should be able to:
> - describe its appearance
> - explain how it was formed
> - give a located example
> - draw a diagram

Diagrams are particularly useful. They do not need to be works of art, but clear, neat and labelled or annotated appropriately in response to the demand of the question. A *label* will simply identify a characteristic of the landform, for example the steep slope of a valley side in the upper course of a river, while an *annotation* may well be needed to explain the characteristic, for example the steep slope of a valley side in the upper course of a river *due to vertical erosion*. Most textbooks will provide you with all you need for this, although you may also undertake some fieldwork in a river environment.

> **Case study tip** You should be able to name and locate an example of each of the landforms listed. They are not detailed case studies, but a small amount of specific information about the location is useful. It could be a specific rock type, some indication of the dimensions of the landform, or perhaps the name of a tributary or a settlement.

You should be aware of some of the factors that influence the rates of these processes (see Figure 11).

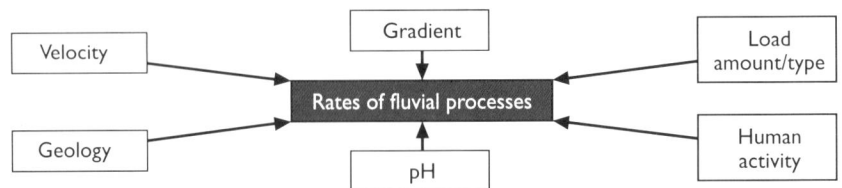

Figure 11 *Factors that influence the rates of fluvial processes*

The relationship between river velocity and process (Hjulström curve)

Figure 12 shows the relationship between the three river processes and velocity for particles of different sizes. You should be able to describe the relationships shown and give reasons for them. Most of this is quite straightforward as it relates to the general principles that:
- river energy increases as velocity increases
- erosion requires more velocity than transportation for particles of the same size
- larger particles have a greater mass, and so require higher velocities for them to be eroded and transported
- deposition occurs sequentially (the largest particles are deposited first) as velocity decreases

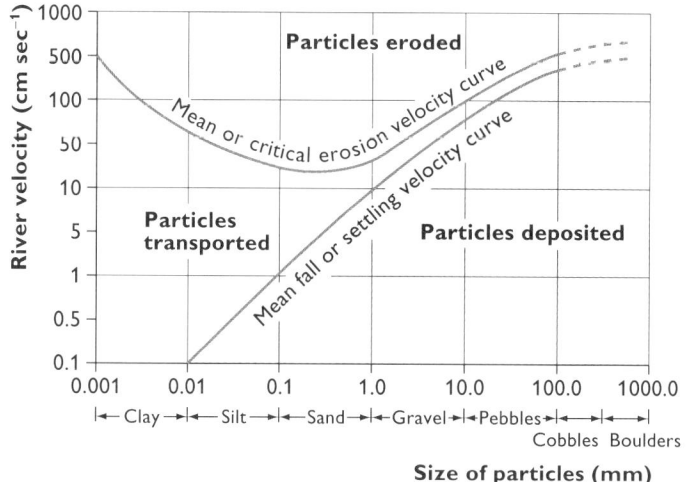

Figure 12 The relationship between river processes and velocity for particles of different sizes

Exam tip There is one odd feature of the graph that is more difficult to explain. The smallest particles (such as clay) require *more* energy for them to be eroded than those that are slightly larger (such as sand). This is because the smallest particles tend to be more cohesive, compacted and electrically bonded. They also have a very small surface area on which the water can exert its energy. Some pretty technical ideas are explained here, but they are well worth trying to remember!

The need for managing river processes and the methods used

This deals with how erosion, transportation and deposition are managed.

Exam tip Do not confuse this topic with managing *flooding*! Some of the methods may be the same, and schemes on some rivers may address both issues, but the emphasis here is on *processes*.

The specific reasons for management and the methods used will vary from scheme to scheme, and you must ensure that you know the case study you have been taught.
- The **need** may be to prevent bank erosion, and the **method** may be the use of concrete mattress revetments.
- Alternatively, the **need** may be to encourage vertical erosion in the centre of the channel in order to maintain an adequate depth for navigation, and the **method** may be the use of wing dykes to trap sediment on the edge of the channel and concentrate flow into the centre.

Both of these examples apply on the Mississippi, but you may well use another river.

Case study tip You must have reasonably detailed knowledge of one scheme. Make sure you know:
- the need
- the method

AS Geography

- how the process is affected
- located detail

A map or a diagram might well be useful. You could then annotate it to fit the demands of the question.

Coastal environments

Coastal processes

The processes of marine erosion, transportation (longshore drift) and deposition

The basic principles of these processes are similar to those of rivers, given that they both involve moving water! However, whereas river processes are due to the continual flow of water, marine processes are largely the result of wave action, and different forces are involved. The mechanisms differ as a result.

Process	Mechanism	What happens?
Erosion	Abrasion	The load of the sea is thrown against the rocks of the coast by breaking waves and wears them away
	Attrition	The particles of load become worn away by colliding with each other and as they hit the coastal rocks
	Hydraulic action	When waves break against a cliff face, air in cracks and crevices becomes compressed; as the wave recedes, the pressure is released, the air suddenly expands and the crack is widened
	Pounding	The force of a breaking wave exerts pressure on the rock, causing it to weaken, even without any material to wear it away (abrasion)
Transportation	Longshore drift	Waves approach the coast at an angle, due to the direction of the dominant wind. After breaking, the swash carries particles diagonally up the beach. Under gravity, the backwash moves them perpendicularly back down the beach. The net result is a movement of material along the beach
Deposition	Loss of energy due to a decrease in velocity	This occurs at the top of the swash and immediately after the breakpoint (which may be offshore)
	Loss of energy due to a decrease in volume	Water percolates into the beach material during the swash and/or backwash, leaving particles on the surface

The material deposited by waves will tend to be smooth and rounded, due to the action of water. The largest material will often be deposited at the top of the beach in storm conditions, when the waves have more energy. Longshore drift will generally move smaller material further along the beach as it is lighter and requires less energy to be moved.

Exam tip If you are *explaining* the process of longshore drift, you would be expected to refer to the gravitational force acting on the backwash.

Factors influencing the rate and location of these processes

The rate (how fast) and the location (where) of the processes depends upon the interaction of a range of factors. These factors vary **spatially** (from place to place) and **temporally** (over time) in their influence and their relative importance. Figure 13 provides a useful summary.

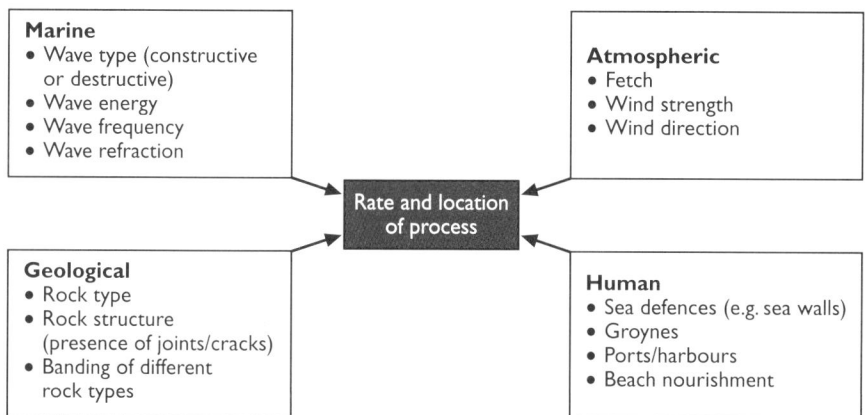

Figure 13 Factors influencing the rate and location of process

Landforms of coastal erosion

It is helpful to deal with these landforms in groups, i.e. a **cliff** and a **wave-cut platform** usually form together, as do a **bay** and a **headland**, while a **cave**, an **arch**, a **stack** and a **stump** form in a sequence.

For each of the landforms you should be able to:
- describe its appearance
- explain how it was formed
- give a located example
- draw a diagram

Most textbooks will provide you with all you need for this, although you may also undertake some fieldwork in a coastal environment. Field sketches, such as Figure 14 (page 34), are also useful, particularly if they are effectively *annotated*.

Case study tip You should be able to name and locate an example of each of the landforms listed. Detailed case studies are not required, but a small amount of specific information about the location is useful. It could be a specific rock type, some indication of the dimensions of the landform, or perhaps the name of a coastal resort.

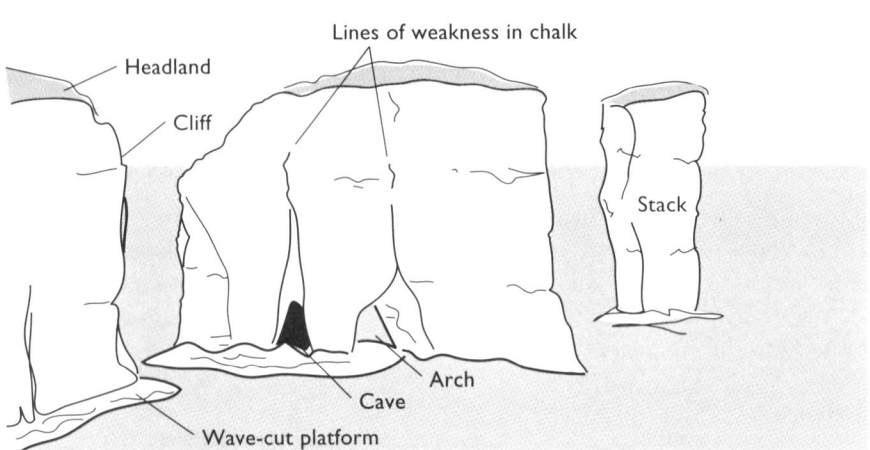

Figure 14 Field sketch of Old Harry Rocks, Dorset

Landforms of coastal deposition

For each of the landforms **beach**, **spit**, **onshore and offshore bars**, **tombolo** and **cuspate foreland**, you should be able to:
- describe its appearance
- explain how it was formed
- give a located example
- draw a diagram

Some of these landforms are better known than others. In particular, cuspate forelands can be problematic! You should be aware that there are different theories about how landforms such as this are formed, and you are therefore likely to be *suggesting* how it *might* have been formed. Again, your textbook should provide you with the detail that you need.

> **Case study tip** As with landforms of erosion, you should be able to name and locate an example of each of the depositional landforms listed. They are not detailed case studies, but a small amount of specific information about the location is useful. It could be a particular rock type, some indication of the dimensions of the landform, or perhaps the name of a coastal settlement.

Coastal landforms

The possible physical and human causes of long-term sea level change

The emphasis in this section is clearly on long-term change, so you do not need to know anything about tidal regimes or to differentiate between the effects of waves at high and low tide.

Eustatic changes are:
- changes in the amount of water in the world's oceans
- global in scale

- caused by physical factors such as global climate change, leading to more/less water being stored as snow and ice on the land; *or* human factors, such as possible changes in the greenhouse effect due to burning fossil fuels

Isostatic changes are:
- changes in the height of the land, causing a relative sea level change
- local in scale
- caused by physical factors such as uplift due to plate collision, or sinking/rising of the crust as the weight of ice sheets is added/removed during a glacial period

> **Exam tip** You should appreciate that changes in relative sea level may be the result of a combination of the two types above. You should also be aware that sea level has changed in the past and may well change in the future.

Landforms of submergence and emergence

Landforms of **submergence** are caused by a relative rise in sea level, the causes of which are detailed above. You should know the terms **ria** (a submerged river valley) and **fjord** (a submerged glacial valley).

Landforms of **emergence** are formed as a result of a relative fall in sea level. **Raised beaches** and **abandoned cliffs** are often produced together as a pair of landforms.

For each of these landforms, you should be able to:
- describe its appearance
- explain how it was formed
- give a located example
- draw a diagram or sketch map

> **Case study tip** As with landforms of erosion and deposition, you should be able to name and locate an example of each of the landforms listed. Detailed case studies are not required, but a small amount of specific information about the location is useful. It could be a particular rock type, some indication of the dimensions of the landform, or perhaps some idea of *when* the landform was formed.

The impact of rising sea level on human use of the coastline

You should notice here that you are only concerned with *rising*, not falling, sea level. There are numerous human uses that could be relevant, and examples chosen to illustrate them may deal with actual or potential impacts. Human uses include:
- tourism
- ports/harbours
- settlement
- agriculture
- industry

> **Case study tip** You should be able to illustrate the specific impact. For example, rising sea levels are thought to have contributed to high rates of coastal erosion on the Holderness coast, leading to significant loss of land and property and major expenditure on methods of coastal protection. You should be able to give brief details

AS Geography

of some of the methods employed and have some idea of the costs involved. The names of villages particularly affected would be helpful.

Coastal ecosystems

Plant succession in a sand dune ecosystem (psammosere) and a salt marsh ecosystem (halosere)

This is a potentially difficult topic and those studying biology may well have a more secure grasp of the concept! However, the principles may have been established on your GCSE course and what you did there should not be forgotten. The key is that the vegetation community changes over time. It can be helpful to see the changes across a transect (see Figures 15 and 16).

	Foredunes	Mobile dune (yellow)	Fixed dune (grey)	Dune slack (fresh water pool)	Dune heath
% bare ground	100	85	10	na	0
No. of different species	1	5	15	27	12
Most dominant (%) Other common species	Sea couch grass (1)	Marram grass (90) Red fescue	Sand sedge (75) Marram, other grasses, creeping, willow and dewberry	Rushes (50) Reeds, grasses, mosses, willow, irises and orchids	Ling heather (90) Heath rush, gorse, broom, buckthorn (some pine and birch)

Figure 15 Transect across a sand dune ecosystem

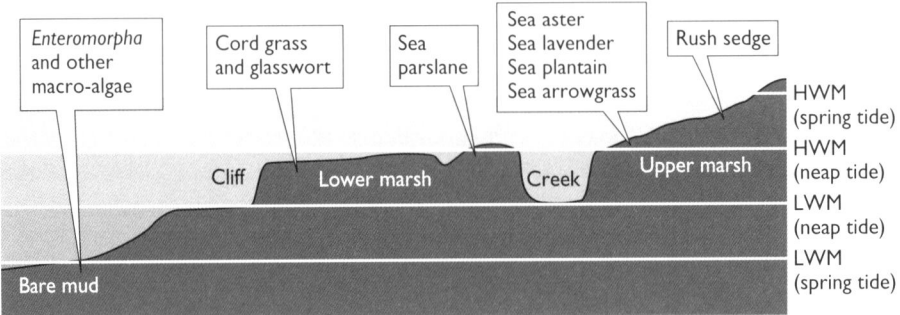

Figure 16 Transect across a salt marsh ecosystem

You need to be able to *describe* the changes over time. During succession, the vegetation tends to become:
- more diverse
- taller
- denser
- more woody in character

You should be able to illustrate these features with detail from your located case study. You also need to *explain* these changes. They occur because over time:
- environmental conditions become less restrictive, for instance when new dunes develop on the seaward side and the older dunes are then protected from salt spray and wind
- soil increases in depth fertility, as more organic matter is added when the vegetation becomes denser

Eventually, a climax community may become established, with particular tree species becoming dominant. This tends to result in *less* diversity as the trees shade out other smaller, lesser species.

> **Case study tip** You must have a located example of each type of ecosystem. Do not confuse the two! Useful detail can be provided by mentioning specific plant species that are found and giving some idea of percentage cover, pH values or vegetation height. You may find it useful to display the changes by means of a cross-section (transect) on which the species are indicated. However, you must ensure that you show by annotation on the diagram or by written text that the changes have happened over time.

> **Exam tip** It is *vital* that you avoid confusing these two types of ecosystem! If a question refers to a salt marsh and you write about a sand dune, you are likely to score no marks at all.

How and why coastal ecosystems are modified by human activity

This may form a useful addition to the examples used above, as succession itself may be affected by human activity.

Modification may be short or long term and deliberate or accidental. Depending on the example you use, it might be the result of:
- tourist impact
- management of tourist impact by the landowners
- protection of the ecosystem, perhaps as a nature reserve

You must be able to say *how* it has been modified as well as *why*. Modification might involve:
- loss of species
- addition of species
- reduction in percentage cover
- increased soil erosion
- loss of wildlife habitat
- scenic damage

> **Case study tip** The key is to support your *how* and *why* comments with some located detail as evidence. This might include the name of the landowner(s), species names, the average number of visitors per year, etc. You only need one located example and if you use a sand dune or salt marsh, you can use the same species detail as for the previous section on succession!

&

This section contains six Unit 1-style questions, two for each of the three sections outlined in the Content Guidance section. Remember that in the unit test you will also have a choice of two questions in each section; you have to answer one of them in each case.

Two model answers are provided after each question. The first is of a typical grade-C standard (candidate A) and the second is of a good grade-A standard (candidate B).

Examiner's comments

The answers are interspersed by examiner's comments (preceded by the icon). These show where credit is due and where errors have been made or improvements are needed. They highlight problems such as irrelevancy, lack of clarity, lack of focus on the wording of the question or shortage of case study detail.

The comments indicate how each answer would have been marked in an exam. In questions only worth a few marks, the examiners may be instructed to give a mark for each relevant point that the candidate makes (up to the maximum). In higher-mark questions, there may be a mark or two for a 'basic' answer (such as the naming of an appropriate process), with additional marks available for 'development' (describing the mechanism of the process). In the final part of the question (usually worth 6 marks), the examiners use a system of 'levels'. These are defined as follows:

Level 1: 1–2 marks
The answer offers simple ideas or a list of points, not all of which may be relevant. A located example (if requested) might not be given.

Level 2: 3–4 marks
Some knowledge and understanding is evident, but cause–effect links may be unclear or simply stated. There may be a lack of range of ideas/information where this is required. Located detail may be unconvincing.

Level 3: 5–6 marks
The answer reveals accurate knowledge and clear understanding. Explanations are explicitly stated with cause–effect ideas well argued. Terminology is used accurately and located detail provided.

Question 1

Earth systems (I)

Study the figure below, which shows the global pattern of tectonic plates.

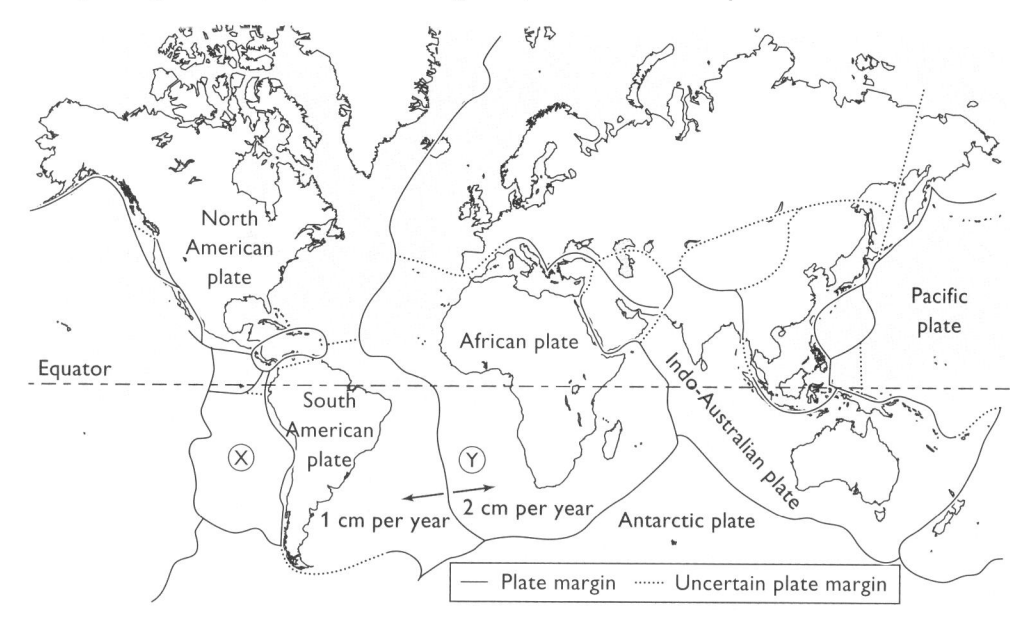

(a) (i) Name plate **X**. (1 mark)
   (ii) In what direction is plate **X** moving? (1 mark)
   (iii) Describe the movement of the plates at margin **Y**. (2 marks)
(b) (i) Suggest why tectonic plates move. (3 marks)
   (ii) Explain how these movements can lead to earthquakes occurring. (3 marks)
(c) Outline the economic benefits of igneous activity. (4 marks)
(d) With reference to a located example, account for the landforms resulting from igneous activity. (6 marks)

Total: 20 marks

■ ■ ■

Answer to question 1: candidate A

(a) (i) Nazca plate

   ✎ This is correct, for 1 mark.

   (ii) North-west

   ✎ This would have been correct for the Pacific plate but it is wrong for the Nazca, which is moving east.

   (iii) The plates are moving apart.

AS Geography

question

e This is sufficient for 1 mark, but the candidate has not added that they are moving at different speeds.

(b) (i) The plates move because they are solid crust and they float on the molten magma in the mantle beneath. There are convection currents in the mantle.

e The candidate has the basic ideas here, and gains 2 marks. However, the answer fails to make the link between the convection cell currents and the plate movement very clear. Another relevant point is that the convection cell currents are generated by heat from radioactive decay in the core.

(ii) When plates move, the ground shakes and vibrates. If these movements are sudden and major, then they are known as earthquakes.

e The candidate shows a very limited grasp here and is awarded only 1 mark. Some of what is written is not strictly true. The key point is that earthquakes occur when **pressure which has built up is suddenly released**. The pressure is due to the friction between the plates as they are moving in relation to each other.

(c) Igneous activity can be of economic benefit because of tourism, valuable minerals and building materials, such as granite which is resistant and durable. It has been widely used in Aberdeen in Scotland which is often known as 'Granite City'. Buildings such as the town hall are built of granite.

e This is a rather mixed answer. The command word is *outline* and so no great detail is actually required and therefore some of the information about Aberdeen is superfluous. Conversely, the first two points need rather more. It should be made clear that tourism creates tertiary employment, for instance; that is why it is an economic benefit. This answer is worth 3 marks.

(d) Located example: Dartmoor

Dartmoor is an area of high relief, which was formed when molten magma rose towards the surface but cooled underground before it reached the surface. This caused the surface rocks to dome upwards. When the overlying rocks were removed by weathering and erosion, the granite was exposed to leave tors on the hill tops.

e This is a moderate answer which is on the right lines but lacks detail and accuracy in the explanation. The range of features covered is rather limited and there is no convincing locational detail. It is a middle-band response worth 3 marks out of 6.

e **Overall, candidate A scores 11 out of 20 marks, which roughly equates to a very low C grade. There is some evidence of sound knowledge, but the depth and accuracy of explanation are rather limited. In one part, the command word is not focused upon. Finally, there is a lack of locational detail in part (d), in which case study knowledge is important.**

Answer to question 1: candidate B

(a) (i) Nazca

e Correct, for 1 mark.

(ii) East

e Correct, for 1 mark.

(iii) The plates are diverging (moving apart) at a relative speed of 3 cm per year.

e This is worth 2 marks, as data have been used from the resource to support the answer. It would, however, have been better if the candidate had given the speeds of each plate instead.

(b) (i) Plates are moved by convection cell currents in the asthenosphere, caused by heat from the core from radioactive decay. The moving magma of the convection cells pulls the crustal plates of the lithosphere with it as it moves.

e This is a good answer, worth 3 marks. The cause is clearly linked to the movement and there is good use of terminology.

(ii) When plates move past each other, friction occurs due to the roughness of the crustal material. Over time, pressure builds up until eventually the pressure exceeds the friction and the plates move past each other. The energy released causes shock waves which lead to vibration and movement of the crust as earthquakes.

e This is a fully explained answer. Although the candidate only seems to be referring to conservative margins, the processes involved are clearly well understood. This is worth 3 marks.

(c) Igneous activity can produce volcanic activity which attracts tourists who spend money in the local economy. In Iceland, geothermal heat is used for domestic heating and industrial power. Minerals such as tin and lead can be mined from areas of granite, as happened in Cornwall. This provided primary employment.

e This is a good answer, worth the full 4 marks. It is not *point-marked* and so you do not need to give four different benefits. There are clear links to the economy and the use of brief examples supports the ideas with evidence.

(d) Located example: Isle of Arran

Rising magma cooled and solidified slowly beneath the surface, forming a batholith of granite with large crystals. This caused the doming of the overlying rocks which were gradually removed by weathering and erosion. Tors are areas of exposed granite which had been subjected to deep chemical weathering before they were exposed. Sills are horizontal outcrops, and dykes vertical outcrops of granite, which were formed when magma seeped from the batholith but still did not reach the surface before solidifying.

question

e This is a good answer, worth 5 marks out of 6. A range of landforms has been dealt with and the explanations are clear and quite accurate for an AS response. A link could have been made between dykes/sills and bedding planes, while some located detail (Goat Fell as an example of a tor, perhaps) should have been offered.

e **Overall, this is a high-quality answer with plenty of evidence of clear understanding. Responses are written accurately with a good grasp of terminology. The command words are addressed and plenty of knowledge has been shown. The candidate gains a total of 19 marks out of 20 — the equivalent of a very high A grade.**

Question 2

Earth systems (II)

Study the figure below, which shows the relationship between rates of chemical weathering and climatic conditions.

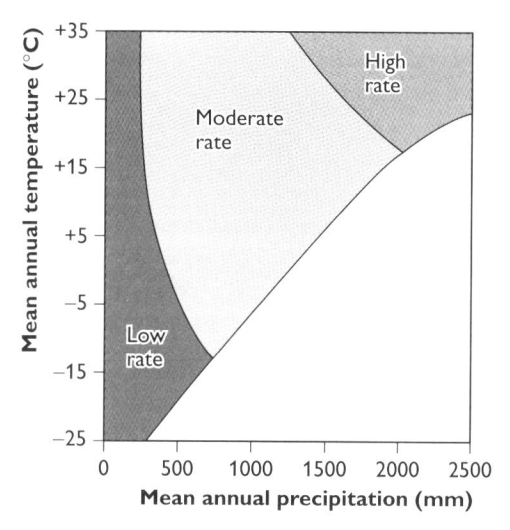

(a) (i) State the minimum temperature associated with high rates of chemical weathering. (1 mark)
   (ii) State the maximum precipitation associated with low rates of chemical weathering. (1 mark)
   (iii) Describe the relationship between rates of chemical weathering and mean annual temperature. (3 marks)
   (iv) Explain this relationship. (3 marks)
(b) (i) Outline the evidence that supports the theory of plate tectonics. (3 marks)
   (ii) Suggest why tectonic plates move. (3 marks)
(c) Describe and explain the occurrence of hot spots. (6 marks)

Total: 20 marks

■ ■ ■

Answer to question 2: candidate A

(a) (i) 22°C

> 🖉 This is incorrect. It would appear that the candidate has read from the edge of the high rate shaded area where it meets the axis rather than at its lowest point at about 2000 mm.

(ii) 250 mm

> 🖉 Again, this is incorrect and it looks as if a similar mistake has been made. The first couple of marks in these questions should be relatively easy to obtain with *careful* use of the resource.

45

AS Geography

question 2

(iii) As the mean annual temperature increases, so too does the rate of weathering. However, it is not a perfect relationship.

e The candidate has recognised a general trend in the relationship, but in fact it is *far* from perfect. Low rates, for instance, occur at all temperatures in the range. The relationship is much stronger with precipitation. It always helps to extract data from a resource to support description. Only 1 mark is awarded here.

(iv) Chemical weathering happens at a higher rate with higher temperatures because the chemical reactions are speeded up. This happens with hydrolysis when hydrogen in rain water reacts with feldspar in granite.

e The answer offers a valid general principle and it is supported with some evidence. In fact weathering is approximately 2.5 times faster if temperatures are increased by 10°C. However, only 2 marks out of 3 are gained because there is no explanation of why the relationship is *not* perfect.

(b) (i) The shape of the continents suggests that they might have been joined together, for example South America and Africa. Fossils of mesosaurus are only found in Africa and South America, suggesting they were at one stage joined. Coal is found in Britain and in Antarctica, while glaciation once occurred in what are now hot deserts.

e Two good pieces of evidence have been provided, although they refer to continental drift rather than to plate tectonic theory. They have been clearly linked to the possible movement. The final piece, however, is not related to the movement at all. Examiners, perhaps a little unkindly, might write 'so what?' after such a statement. The evidence should be linked to the theory more closely. This is worth 2 marks.

(ii) The plates are moved by convection currents in the mantle. The plates 'float' on top of the molten magma in the mantle and the convection cell currents pull them along as they spread out beneath the surface.

e This is a pretty sound response at this level and is worth 2 out of 3 marks. It could have been improved by explaining *why* convection currents occur and/or providing a slightly more accurate explanation as to the link between the currents and the plates, perhaps referring to the lithosphere.

(c) Hot spots are found in places on the centre of plates away from the plate margins. Most volcanic activity takes place at the margins but it can also happen in the centre of the plate at a hot spot.

At a hot spot, there is rising magma coming up from the mantle. This is very hot and under a great deal of pressure. It forces its way through the plate and emerges on the surface as a volcano. The lava is runny and so it flows a long way, giving gently-sloping sides to the volcano cone.

e The candidate is clearly familiar with the term 'hot spots' and has a reasonable idea of their existence. Some of the terminology and language is a little inaccurate and

Edexcel (A) Unit 1

there is some repetition in the first couple of sentences. It is worth thinking about the structure of your answer to these longer parts of the question, even if only for a few moments, in order to gather your thoughts. The response would also have benefited from the use of an example, even though the question did not require this. It is a middle-level answer worth 4 marks.

The total mark for this answer is 11 out of 20, which generally equates to a reasonably sound C grade. It was a shame that some relatively accessible marks were dropped at the start as there were signs of a clear grasp of most of the topics. A little more care and a little more thought given to planning would both have helped.

Answer to question 2: candidate B

(a) (i) 18

 This is correct, for 1 mark, and has clearly been read carefully from the resource. The units should really be quoted, but this does not deny the candidate the mark as the skill has been demonstrated.

(ii) 750 mm

 This is also correct, for 1 mark. The figure is within the acceptable tolerance limits that the mark scheme always allows when values are hard to read accurately from a graph with few or no grid lines.

(iii) The general relationship is that higher rates of weathering occur at higher mean annual temperatures. High rates, for example, only occur at temperatures over 18°C and moderate rates when temperatures exceed −15°C.

 This is worth 2 marks as the general trend has been identified and good use has been made of supporting data as evidence. No reference has been made to the imperfect nature of the relationship and so 1 mark is lost.

(iv) Chemical reactions usually happen faster in higher temperatures and so this will speed up weathering rates. They are at their highest in tropical rainforest areas where there is high rainfall, giving water which is needed in most chemical processes, and also high temperatures over 20° to speed up the reaction.

 There is some sound explanation here and the tropical rainforest illustration helps. The point could have been further developed, however, to appreciate that rates are low in arid tropical areas despite the high temperatures. The availability of moisture is therefore very important too. The answer is worth 2 marks as it stands.

(b) (i) Studies of the mid-Atlantic sea floor have revealed that basalts are older the further away they are from the ridge, suggesting they were formed at the ridge and have then been moved away. The rocks also have parallel bands of magnetism, alternately N and S, forming a symmetrical pattern either side of the ridge, indicating that they were formed together at the centre. Continental shelves also seem to fit together like a jigsaw.

> *e* This answer takes a different approach to that of candidate A and there is a clear focus on plate tectonic theory. The evidence is well explained, providing more than enough detail for an 'outline'. Three valid ideas have been clearly presented and the full 3 marks are awarded.

(ii) The lithosphere (crust and solid mantle) floats on the molten magma of the inner mantle. Convection cell currents in the mantle, caused by radioactive decay in the core, pull the plates apart or push them together.

> *e* This is a good answer at this level. There is clear understanding of the structure of the earth, and a reason for the existence of the convection currents is offered. The 'push' and 'pull' may be a little simply stated, but this does not detract from the overall quality of the answer. All 3 marks are awarded.

(c) Hot spots are areas of active volcanic activity lying well away from the margins of tectonic plates. They may occur under oceans, giving rise to volcanoes such as Mauna Loa on Hawaii, or on the continental land masses, such as the geysers in Yellowstone, USA. They tend to remain static for a long period of time and the plates sometimes move over the top of them, changing the position of the volcanic eruption.

> *e* This gives a good answer to *part* of the question. It is an effective description of hot spots, illustrated with contrasting examples. The question also demands explanation, however, and this has been neglected. There needs to be reference to the existence of rising plumes of magma from the mantle, exploiting thin/weak crust. As a partial answer it only reaches the middle level and is worth 4 marks out of 6.

> *e* **Candidate B scores a total of 16 marks out of the 20 available for this question. This would be the equivalent of a safe A grade. There are some well-written answers with good use of terminology, and effective application of illustrations and supporting evidence. However, there was insufficient focus on both parts of the question (c); candidates should always look for the use of double command words, especially in the longer, final part.**

Question 3

Fluvial environments (I)

Study the figure below, which shows the regime of a river in North America.

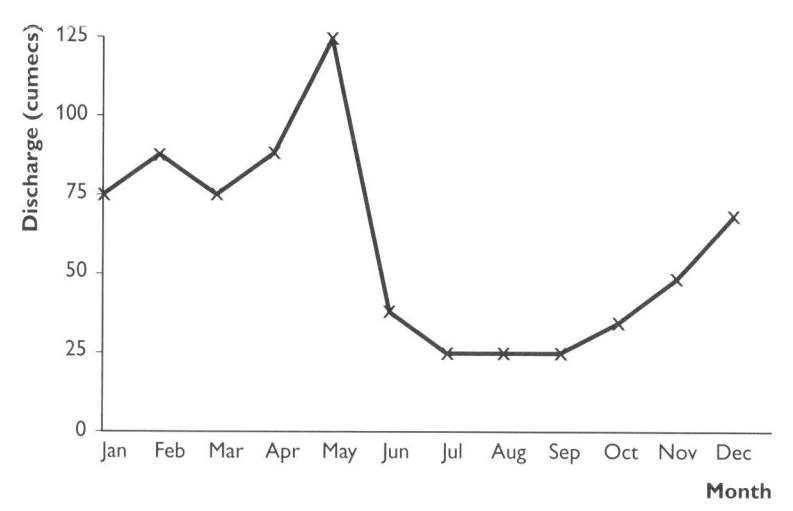

(a) (i) Estimate the range of discharge values on the graph. (1 mark)
   (ii) Describe the regime of the river. (3 marks)
   (iii) Suggest reasons for the regime of the river. (4 marks)
(b) (i) Describe the processes by which a river erodes its channel. (3 marks)
   (ii) Outline the influence of the factors affecting rates of river channel erosion. (3 marks)
(c) With reference to a located example, describe how and why a river process may be managed. (6 marks)

Total: 20 marks

Answer to question 3: candidate A

(a) (i) 100 cumecs

   e Correct, for 1 mark. This is obtained by subtracting the minimum discharge (around 25 cumecs) from the maximum discharge (around 125 cumecs).

   (ii) It starts off fairly high in January and then rises rapidly in about May to a peak. It then falls quite quickly in the summer before rising again in October.

   e This is a description of the changes in discharge but there is no use of data to support the description, nor is there an appreciation of the general trend. Only 1 mark is awarded.

   (iii) There is a higher discharge in the winter than in the summer, probably due to

49

higher rainfall totals. In the summer, the temperatures are higher and so there will be more evapotranspiration, meaning less water reaches the river. When the ground is dry in the summer it is able to store water in underground stores and in the soil. The rapid rise is probably because the relief is steep and the rock type impermeable, and so rainfall quickly reaches the river, giving a high discharge.

e There are several valid factors here and they are well linked to the regime pattern. The obvious one that has been overlooked is the likely spring snow melt which causes the rapid rise in May. As this is a key feature of the regime, only 3 out of 4 marks are awarded.

(b) (i) Abrasion is when the load of the river rubs against the bed and banks, wearing them away. Attrition occurs when the load itself becomes smaller as it is eroded by colliding with other particles and with the bed and banks. Hydraulic action is the impact of the water wearing away the banks.

e Abrasion is outlined successfully, but attrition does not erode the channel (a lack of focus on the precise wording of the question) and hydraulic action is not correctly outlined. This answer is worth just 1 mark out of 3.

(ii) How fast the river is flowing is a major factor as is the type of rock it flows over. The load is also important. It may be reduced by human activity.

e This amounts to nothing more than a list, gaining only 1 mark. It is a shame, as the factors are relevant, but their influence must be made clear. Again, there is a lack of focus on the question — this time, the command word.

(c) Located example: Mississippi

Huge lengths of concrete mattress revetments have been laid on the river bank by the Corps of Engineers. This is unsightly but it is effective. Wing dykes have been built out into the channel from one bank and this has trapped sediment. It also forces the water into the centre of the channel and this concentration of flow has caused more erosion of the bed, making the channel deeper.

e This is a pretty good answer to half of the question. The candidate has focused on *how* the river processes have been managed and two appropriate methods have been discussed. There is limited locational detail, but at least the Corps is mentioned to provide something specific to the area. Sadly, there is no reference to *why* the processes are managed in these two ways. This is a middle-level answer, for 4 marks.

e **Overall, candidate A appears to have slipped up by not reading the question carefully, scoring a total of only 11 out of 20 marks. There is evidence of some wide-ranging knowledge and sound understanding, but marks have been lost in three parts by a lack of focus on the specific demands of the question.**

Answer to question 3: candidate B

(a) (i) 100 cumecs

   ✎ Correct, for 1 mark.

   (ii) Overall, the regime shows a high level of discharge in the winter and a significantly lower level in the summer. There is an exception to this general pattern and that occurs in May when there is a sudden and rapid rise to a high peak.

   ✎ Only 2 marks out of 3 are gained here for identifying the general trend and the anomaly. However, no data have been used.

   (iii) There is a high discharge in the winter compared with the summer, probably due to higher rainfall totals being received. In the summer, the temperatures are higher and so there will be more evapotranspiration and so less water reaching the river. The rapid rise in May is likely to be the result of spring snow melt which appears to reach the river very quickly, suggesting a lack of vegetation cover, steep slopes and an impermeable geology.

   ✎ This is a full, clear explanation with appropriate reasons, worth all 4 marks.

(b) (i) The channel is eroded by abrasion (or corrasion) when the river's load rubs against the bed and banks, wearing them away. Corrosion (or solution) may also be important if the water is weakly acidic and dissolves certain minerals in the rock. Cavitation is the combined force of hundreds of bubbles bursting and exerting pressure.

   ✎ The processes are all valid and clearly outlined. Cavitation is explained a little simplistically, but at this level and in response to the command *describe*, it is good enough and 3 marks are awarded.

   (ii) The rate of erosion is controlled by the power of the water and the resistance of the rock of the bed and banks. Steeply-sloping, fast-flowing rivers with lots of load are able to erode rapidly due to their excess energy and tools of erosion. If the bedrock is weak (clay), then erosion is also rapid. It is slower on resistant rock (e.g. granite).

   ✎ The candidate makes an interesting general statement at the start, suggesting an understanding of the principles. The role of geology is clearly addressed and the river variables are sufficiently well outlined for a full 3 marks.

(c) Located example: Mississippi

In order to maintain a 9 m depth of water for navigation, wing dykes were built on the banks. These trapped sediment at the edge of the channel by slowing velocity and causing deposition, while increasing erosion in the centre, making it deeper.

After meanders had been straightened, the river tried to meander again. To stop erosion on the outside of the meander bends, rip rap (boulders) and concrete revetments were laid on the river banks south of St Louis. This reduced erosion and maintained the straight channel.

e The candidate has addressed *how* and *why* and included some located facts and figures. This is a well-focused response, worth 6 marks.

e **Overall, candidate B scores 19 out of 20 marks which equates to a high A grade. Some of the explanation is full and detailed and there is good focus on the wording of the question. There are weaknesses in some parts, but at AS answers do not need to be perfect! Full marks can be gained if you provide enough correct material, even if mistakes are also made.**

Question 4

Fluvial environments (II)

Study the figure below, which shows a drainage basin cycle.

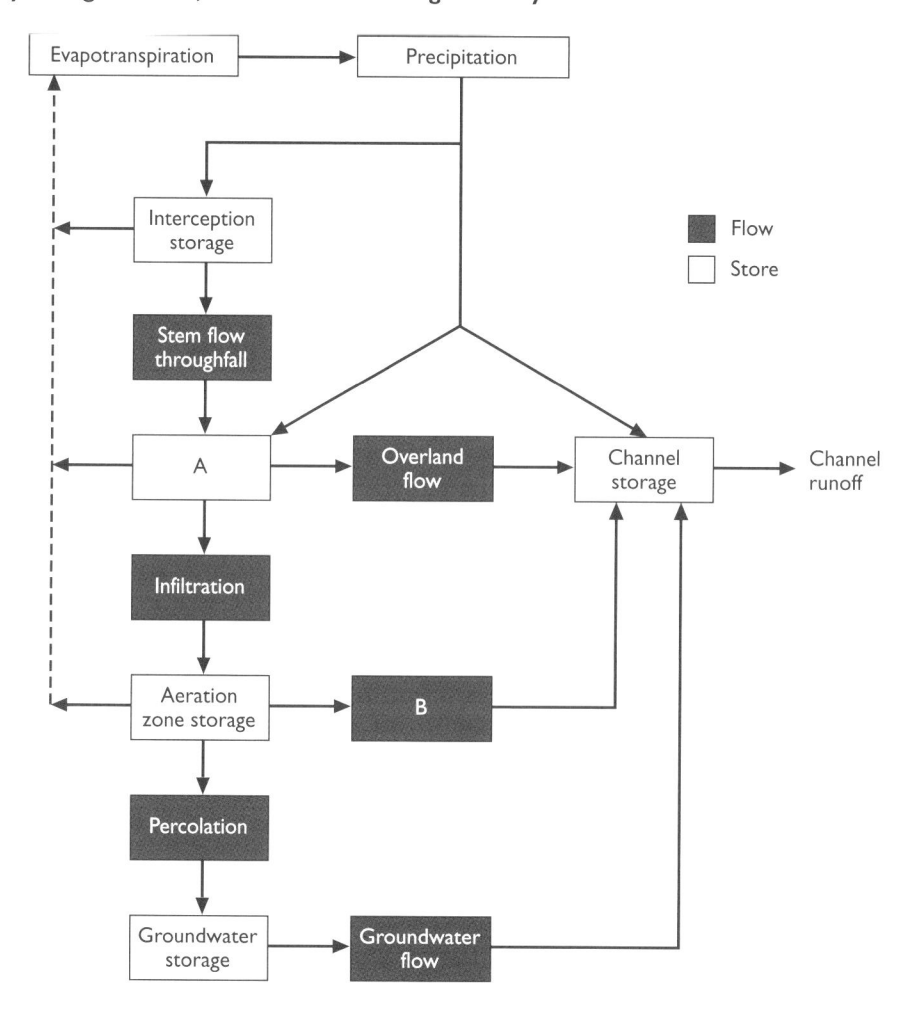

[Diagram: Drainage basin cycle showing Evapotranspiration ↔ Precipitation, leading to Interception storage, Stem flow throughfall, store A, Overland flow, Channel storage, Channel runoff, Infiltration, Aeration zone storage, flow B, Percolation, Groundwater storage, Groundwater flow. Key: Flow (shaded), Store (unshaded).]

(a) (i) Identify store **A** and flow **B**. (2 marks)
    (ii) Define the term **evapotranspiration**. (2 marks)
(b) Explain why interception storage might vary:
    (i) from basin to basin (3 marks)
    (ii) over time (3 marks)
(c) Outline the causes of rainfall at frontal systems. (4 marks)
(d) With reference to a located example, explain the physical causes of flooding. (6 marks)

Total: 20 marks

Answer to question 4: candidate A

(a) (i) A = puddles; B = soil water

e Neither of these answers is correct. A should be surface storage or depression storage. Puddles are only one element of this. B is throughflow, which is the movement of water laterally through the soil. No marks.

(ii) Evapotranspiration is a combination of evaporation and transpiration from the vegetation.

e This answer gains 1 mark for revealing some knowledge of the process, but it is rather limited. There is some awareness that it may be linked to vegetation, but a definition needs to do more than rewrite the term in question! For full marks, the answer should demonstrate some specific knowledge of the mechanisms involved.

(b) (i) Interception varies from basin to basin because some basins will have grass, some trees and some may be bare.

e This is a very limited response, worth only 1 mark out of 3. The candidate has used time and space to repeat much of the wording in the question, which is unnecessary. Although some relevant reasons have been identified, their influence on interception has not been explained. In addition, there are other reasons, such as the density of vegetation cover which may vary even if the type of cover is the same.

(ii) Interception also varies as the vegetation may be more dense in the summer than in the winter, particularly in deciduous woodland, and so a greater percentage of precipitation will not reach the ground directly but land on the leaves. Agricultural activity can also play a part as crops may be harvested at the end of the summer, leaving less vegetation to intercept rainfall in the autumn.

e This is a much better response and worth the full 3 marks. Two different ideas are offered and there is a reasonable explanation of each of them. The key is to establish a clear cause–effect linkage.

(c) At fronts there is a meeting and mixing of cold and warm air. When warm air approaches cold air it rises, because the warm air is lighter than the cold air. As it rises, it cools down at a lapse rate until condensation occurs, leading to the formation of water droplets and cloud cover. If the water droplets collide with each other and get heavier, then rain can fall.

e This is a reasonable answer and several processes are specifically mentioned. It scores 3 out of 4 marks. It could have been improved by referring to the **low density** of warm air rather than it being lighter. The answer also does not explain why condensation occurs at a particular stage in the sequence, nor does the impact of mixing on temperature receive consideration. The reference to lapse rates is vague, but in fact there is no requirement for candidates to know about lapse rates; it is an A2 topic (Unit 4).

Edexcel (A) Unit 1

(d) Located example: the Yellow River, China

There was a major flood on the Yellow River during 1998 and over 3000 people were killed. Damage was estimated at over £10 billion.

There were two main reasons for the flood. First, there was a long period of heavy rain. This caused the ground to be saturated and the water table to be high. Further heavy rain was unable to infiltrate and so there was much rapid surface runoff causing the river level to rise and eventually flood. The other reason was soil erosion in the upper course due to massive deforestation for agriculture. This caused the river bed to rise by 10 cm a year. Less water can therefore be held in the channel and it overflows its banks more easily.

> This is a rather mixed answer with some good parts but also some irrelevancies. The example is a valid one and the candidate has some located knowledge. The reference to heavy rainfall is sound and there is a clear explanation as to why this leads to flooding. It is a shame that this part is not supported by some rainfall data. The first part about the deaths and damage is irrelevant as the question is about the causes, not the impact. The final idea is interesting and the rising bed is acceptable as a physical cause, even though the main reason for the increased soil erosion is deforestation, which is a human cause. The answer is worth 3 out of 6 marks as a middle-level response.

> **Candidate A is awarded 11 marks out of 20 for this question, which is equivalent to a C grade. There is evidence of both knowledge and understanding and in places there are some well made links between cause and effect. There is a tendency for the answers to contain irrelevancies and the candidate needs to ensure that the very specific demands of each question are focused upon.**

■ ■ ■

Answer to question 4: candidate B

(a) (i) A = surface storage; B = throughflow

> Both parts are correct, for 2 marks.

(ii) Evapotranspiration is the combined process of evaporation, from the ground surface and from leaf surfaces, and transpiration, which is the passage of water through the stomata of leaves into the air.

> This is a sound definition, worth 2 marks. Both elements are recognised and there is clear knowledge of the mechanisms.

(b) (i) Interception is higher if the vegetation cover is more dense, as there is a greater surface area of leaves to intercept the rainfall. Broad-leaved vegetation, such as capoc in tropical rainforests, also gives a greater cover than grass or the needles of coniferous trees.

> *e* This answers the question very clearly and accurately. There are two ideas, each of which is well explained. It gains the full 3 marks.

(ii) During the course of the year, interception varies as the density of leaf cover changes. Deciduous trees shed their leaves in winter and are bare, therefore not intercepting very much at all. Human activity also plays a part. If a natural forest experiences deforestation, then the density of cover and the effectiveness of interception decrease significantly. This can expose the soil to erosion.

> *e* This is another good answer, worth 3 marks. Again, two different ideas are offered and their influence is clearly explained. The final comment is irrelevant, but this does not lead to the loss of a mark; credit is always given to the valid content.

(c) Rain occurs when a warm front meets a cold front. Clouds are produced and rain follows.

> *e* This gains no marks. It would appear that this candidate, who has so far scored well on each sub-section, has a knowledge gap. It is a requirement for candidates to cover all parts of the AS specification and topic selectivity should be avoided. It is possible that the same topic could occur in both questions in a section of the examination paper and so you could not avoid it! Be warned!

(d) Located example: Mississippi River

In 1993, the Mississippi basin was subjected to prolonged rainfall in May and June, which caused the water table to rise and the ground to be saturated. Spring snow melt from the Rockies also added to the problem. In July and August there were heavy storms, with some places having 150% more rainfall than average. This resulted in much surface runoff as both infiltration and percolation were limited by the very wet antecedent conditions. The river rose rapidly and exceeded bank-full, even at places like St Louis, which was protected with levees.

> *e* This answer has a strong focus on rainfall in the basin and it is sufficiently well supported with data and detail. There are clearly explained cause–effect links and effective use of accurate terminology (antecedent conditions, bank-full). It is a top-level answer, worth the full 6 marks. Again, credit is given for what is included and how well it meets the top-level criteria, rather than marks being lost for what has been left out.

> *e* **Overall, this candidate gains 16 marks out of the 20 available, which equates to a good A grade. The answers are generally succinct, accurate and well-explained where required. It is let down only by the gap in knowledge of rainfall processes. This candidate had a good enough depth of understanding for that weakness not to be too costly. A weaker candidate may have paid a greater penalty. Even so, this candidate has missed out on several marks which could make a difference to the final grade achieved if performance is less convincing in other units.**

Question 5

Coastal environments (I)

Study the figure below, which shows the annual transfer of sediment along a stretch of coastline (in '000 m³ per year).

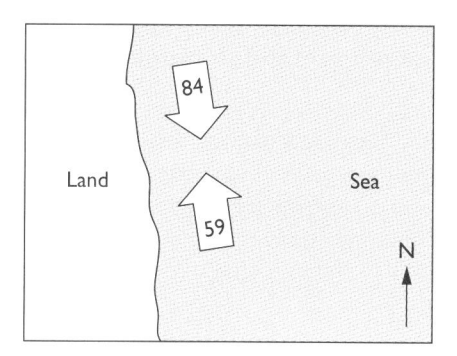

[Diagram showing a coastline with Land on the left and Sea on the right. A downward (southward) arrow labelled 84 and an upward (northward) arrow labelled 59. North arrow points upward.]

(a) (i) Calculate the difference between the amounts of northwards and southwards sediment transfer. (1 mark)
   (ii) From which direction does the dominant wind appear to come? (1 mark)
   (iii) Explain the process by which sediment is moved along a stretch of coastline. (3 marks)
   (iv) State two sources of sediment other than transfers along the coast. (2 marks)
(b) (i) Outline the processes by which the sea erodes the coastline. (3 marks)
   (ii) Suggest why rates of marine erosion vary. (4 marks)
(c) With reference to a located example, describe and explain the plant succession of a psammosere (salt marsh ecosystem). You may use a diagram to assist your answer. (6 marks)

Total: 20 marks

■ ■ ■

Answer to question 5: candidate A

(a) (i) 25 000

   e Correct, for 1 mark, though it would be better to give the units as well.

   (ii) North

   e Correct, for 1 mark. As more sediment is moving southwards, the wind must come from a broadly northerly direction more often than from the south.

   (iii) The process is longshore drift. The wind causes waves to approach the coast at an angle and sediment is moved diagonally up and across the beach. This is swash. The backwash then brings the material straight back down the beach and so there is a net movement of sediment along the beach.

   e The process has been correctly identified and the reason why the material moves

AS Geography

question 5

along the beach has been explained. The response is worth 2 marks rather than 3, however, because there is no explanation for the direction of the backwash.

(iv) Material can come from the erosion and weathering of cliffs as well as from humans.

e This is worth 1 mark for the cliff, but the reference to humans is too vague. In response to the command word *state*, the candidate does not need to write in full sentences.

(b) (i) When waves break, they hit the coast with force and this weakens the rock. This is wave pounding. Material carried in the water, such as pebbles, also rubs against the rocks, wearing them away by abrasion. The pebbles are also worn away by the impact and this is called attrition.

e Only an *outline* of the processes is required and pounding and abrasion are both valid, for 2 out of 3 marks. The question specifically refers to the erosion of the coastlines and so attrition is not relevant.

(ii) Erosion varies because at high tide the waves are larger and have more energy than at low tide and so erode more rapidly. At certain times of the year such as in the spring, tides are higher, and so more erosion occurs.

e This question has been approached from the temporal perspective, i.e. rates varying over time. The spatial view could also have been taken, i.e. from place to place. The answer has a limited range of ideas and the height of the tide does not necessarily mean that waves will be larger. It is a basic-level answer, worth only 2 marks.

(c) Located example: Studland

Near the beach, there are few plants due to the harsh conditions. As you go inland, the amount of vegetation increases and there is a greater variety of species. The vegetation also becomes taller. It starts off with marram grass and then further inland you also find gorse, heather and finally pine trees. It changes because further inland the conditions are not so harsh and so the other species are also able to survive.

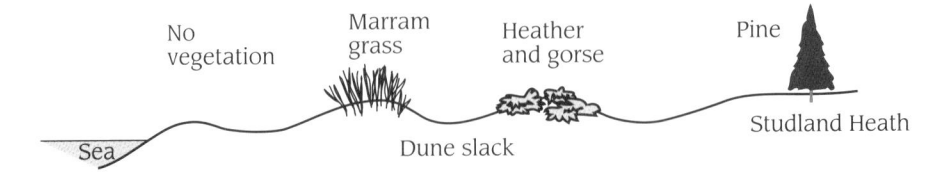

Diagram: Cross-section from Sea showing "No vegetation", then "Marram grass", "Dune slack", "Heather and gorse", and "Pine" at Studland Heath.

e This is a partial answer, which sees succession as a spatial idea, i.e. the vegetation varies across the dunes from the seaward side. The key to succession is that it is a temporal process, i.e. it happens over time. There is some species detail which helps to secure the location, but the depth of explanation is limited. The diagram does not really add anything to the written answer as again it shows a spatial view

Edexcel (A) Unit 1

and there is no annotation to enhance the explanation. The reasons for the conditions being 'harsh' are unclear. It is a middle-level answer, worth just 3 marks.

e **Overall, this response is worth a total of 12 out of 20 marks, which is the equivalent of a firm C grade. There are plenty of signs that the candidate possesses sound knowledge of the topics concerned. When the command is 'explain', some gaps and weaknesses in the depth of the candidate's understanding are revealed.**

Answer to question 5: candidate B

(a) (i) 25 000 cubic metres

e Correct, for 1 mark.

(ii) North-east

e This is a valid answer, for 1 mark, as the wind could be from this direction or due north.

(iii) Because the dominant wind is at an oblique angle to the coastline, waves also approach from the same direction. When the waves break, the swash carries beach material diagonally across the beach. The backwash takes material perpendicularly back down the beach under gravity. Overall, there is a net movement of beach sediment along the beach.

e This is a full, accurate answer with a clear explanation of the process. The fact that the process (longshore drift) has not been named does not matter and so a full 3 marks are awarded.

(iv) Beach nourishment and cliff weathering.

e Both of these ideas are valid, for 2 marks.

(b) (i) Hydraulic action occurs when breaking waves compress air in cracks and crevices on the cliff face. This weakens the rock and enlarges the cracks, leading to undercutting and collapse. Corrasion also takes place as sediment is hurled against the cliff when waves break, wearing away the cliff face. Solution occurs as sea water is acidic and dissolves the rock it comes into contact with.

e There are two valid processes outlined, but sea water is typically neutral and so solution will only really be significant when associated with the secretion of organic acids by, for example, sea urchins. This is worth 2 marks.

(ii) Some parts of the coast are made of weak rock, such as clay, which is eroded rapidly, while more resistant rock, such as granite, erodes more slowly. Wave energy also varies. Some waves are high and break vertically with great energy, while others lack height, break at a low angle and deposit rather than erode.

59

e The first part of this answer is valid and clearly explained. The second part is a little vague and it is not clear whether the candidate is referring to spatial or temporal variation. There is no reason offered as to why waves are higher in some instances. However, it is worth 3 marks because there is development of the first point by the provision of examples of different rock types.

(c) Located example: Studland

Over time, the vegetation changes. It has become taller, denser and more varied. Initially there would only have been a few low plants such as couch grass, which in time is replaced by marram. This traps sand and stabilises the dunes, allowing other, less tolerant species to take root, such as heather and mosses. With more vegetation there is more dead organic matter to aid soil development. Deeper, more fertile soils also allow more species and taller species to develop and, eventually, the climax community of pine woodland develops.

The reason for these changes is that, initially, the conditions are harsh near the sea, due to the wind, salt and high tides. Over time, the conditions improve and this allows the other species to develop.

e This is a good answer and it does not matter that a diagram has not been used. There is a clear description of the changes and there is an emphasis on the temporal element. Some appropriate species detail has been given. The explanation is a little muddled and the candidate has not clearly established why the conditions improve over time (the growth of new dunes on the seaward side). It is a higher-level answer, worth 5 out of 6 marks.

e **Candidate B has produced a competent answer, worth a total of 17 out of 20 marks. In most parts, the depth of explanation is good and this conveys a firm impression that the candidate has a secure understanding of the topics concerned. There is a clear focus on the demands of the question. In a couple of places, there is slight lack of clarity which hampers the response, but enough credit has been gained for a top grade.**

Question 6

Coastal environments (II)

Study the figure below, which is a sketch of a stretch of coastline.

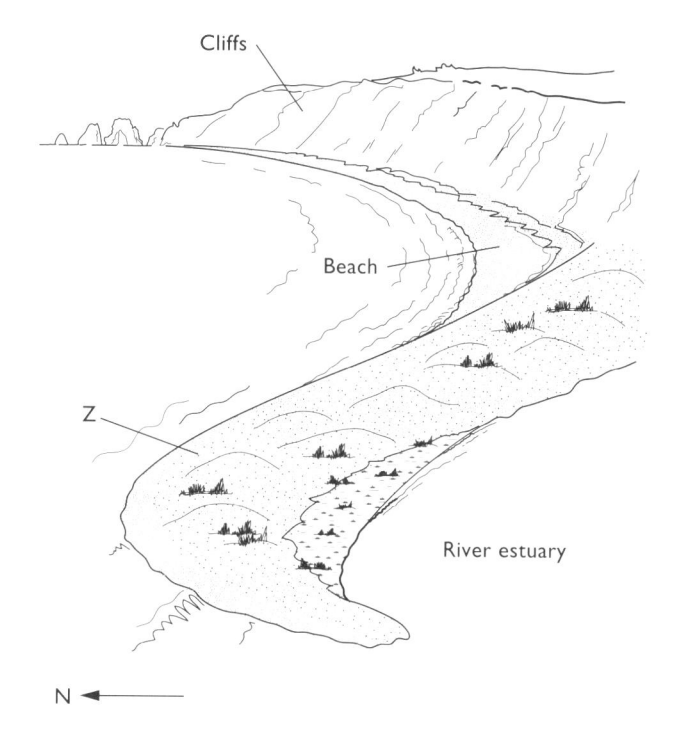

(a) (i)   Name feature Z. (1 mark)
    (ii) Mark and label an area of marsh. (1 mark)
    (iii) From which direction does the dominant wind appear to come? (1 mark)
    (iv) Suggest reasons for the curved shape of the end of feature Z. (3 marks)
(b) (i)   Explain how a relative sea level rise can be influenced by:
        (1) eustatic change
        (2) isostatic change (4 marks)
    (ii) Outline the impact of rising sea level on human use of the coastline. (4 marks)
(c) With reference to a located example, describe how coastal ecosystems are modified by human activity. (6 marks)

Total: 20 marks

Answer to question 6: candidate A

(a) (i) Spit

✎ This is correct, for 1 mark.

(ii)

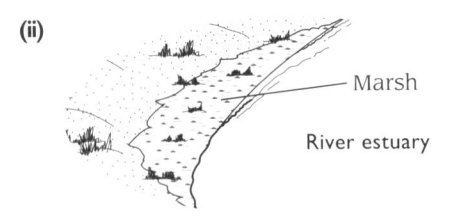

Marsh

River estuary

e This is correctly identified, for 1 mark.

(iii) East

e Correct, for 1 mark. Remember to give wind directions by the direction the wind comes from, not the direction it blows towards.

(iv) It is curved because the waves curve around the end of the spit and push sand around the corner. This is called refraction.

e 1 mark only is awarded for a basic awareness of refraction, which is not fully explained. There are other reasons as well, such as the depth of water, the river current and possibly winds from the south-west.

(b) (i) (1) Eustatic is when the amount of water in the sea increases and so the sea level goes up in relation to the land.

e This is worth 1 mark out of 2 as the principle is correct but there is no explanation of what causes such changes in the amount of water.

(2) Isostatic is when the level of the land goes down and so sea level appears to rise in comparison.

e Again, this is a partial answer, worth 1 mark out of 2. *Why does land level change?*

(ii) A rising sea level can cause severe problems of flooding in flat, low, coastal areas such as East Anglia. This can damage buildings and property and can cost huge amounts of money for insurance companies and residents. In extreme cases, people can be killed if they are washed away by flood waters.

e Flooding is a relevant effect and the candidate has offered some detail about the specific impacts. It is a very narrow view, however. What about agriculture, transport...? The question refers to 'activity' and so some idea of the human response would also be helpful. It gains 2 marks out of 4.

(c) Located example: Studland, Dorset

The sand dune ecosystem at Studland has been severely damaged by the large number of tourists who visit it each year. They drop large amounts of litter and they wear the footpaths away badly. Sometimes they light fires or have barbeques and this can set fire to the vegetation, especially when the grass is very dry in summer. Animals and birds may also be frightened away from their habitat by the noise as well as the fire. The fire may cause a loss of their habitat.

🇪 This is a relevant example to use, but the answer fails to meet some important requirements. There is not much focus on the 'ecosystem'. How does litter and footpath trampling modify the ecosystem? The reference to fire is better, as it is linked to habitats and species. The answer also lacks located detail, such as the names of species. The candidate could be writing about anywhere! It is a middle-level answer, worth 3 out of the 6 marks available.

🇪 In total, candidate A scores 11 marks, equivalent to a C grade. The small number of marks in each of the early sub-sections are successfully picked up and no careless mistakes are made. A moderate level of understanding is evident in part (b), but greater depth is required. When you are 'explaining' something, keep asking yourself 'why?' The final part is sound but could easily have been improved upon with some located detail and a little more focus on the question. This would have lifted a C-grade answer towards an A grade.

■ ■ ■

Answer to question 6: candidate B

(a) (i) A spit

🇪 Correct, for 1 mark.

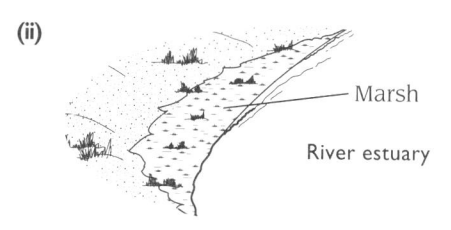

(ii)

[Diagram showing a coastal feature with labels "Marsh" and "River estuary"]

🇪 Correctly marked and labelled, for 1 mark.

(iii) North-east

🇪 Although this is a different answer from that offered by candidate A, it is also valid, as wind from this direction would have the same effect.

(iv) The end of the spit has a curved or hooked shape because the flow of the river estuary will prevent longshore drift from taking sediment any further. It builds up at the end and is carried landwards by the more powerful waves in the deeper water. The south-east waves will refract around the end of the spit, also carrying sediment towards the land.

🇪 This is a full, accurate answer, worth all 3 marks. More than one reason is offered and the explanations are clear and relevant.

(b) (i) (1) Eustatic sea level rise occurs when there is an increase in global temperature causing ice sheets and glaciers to melt. More water in the hydrological cycle is returned to the ocean store, raising sea level.

(2) Isostatic change happens when weight is added to the land surface, perhaps in the form of a developing ice sheet. The land sinks and so the sea level rises in relation to it in that region.

e Both of these are spot on, for a full 4 marks. The explanations are complete and there is also a recognition of the differences in scale (global compared with regional).

(ii) Rising sea level can cause the flooding of settlements, loss of agricultural land and the disruption to communications if bridges and roads are submerged. Tides will gradually become higher and there could be more coastal erosion.

e This is something of a list and although each point made is relevant, there is no exemplification (detail/evidence) and the human response is not addressed. It scores 2 marks out of 4. Only give a list in response to the command 'list' or 'state'.

(c) Located example: Studland sand dunes in Dorset

Studland receives over 1 million visitors per year and they need to be managed. To protect the ecosystem from damage, the National Trust has fenced off certain areas of marram grass so that it is not trampled. Areas that have been trampled and damaged can be replanted. The use of barbeques is restricted to a designated area and boardwalks are provided so that footpaths are not eroded. To prevent erosion and to protect the beach huts, gabions and wooden palisades have been installed, although the National Trust now has a policy of managed retreat, which allows nature to take its course.

e Candidate B clearly has a secure knowledge of the local area and some specific data and detail are provided (number of visitors, National Trust, species name). There is, sadly, a lack of focus on the demands of the question. As with candidate A's answer, there is limited reference to the Ecosystem and rather too much about coastal erosion. It is, therefore, a middle-level answer and scores 3 out of 6 marks.

e **Overall, candidate B scores 15 out of 20, which is a safe A grade. The levels of knowledge and understanding are both impressive and the candidate seems well prepared to answer a question on this topic. However, in two sub-sections, marks are lost by a lack of focus — once on the command word and once on the specific aspect of the topic in question. With better technique, the candidate could have been approaching full marks.**